This Good Earth

This Good Earth

The view from AUDUBON magazine

Edited by Les Line

Published in cooperation with the National Audubon Society by

CROWN PUBLISHERS, INC. : NEW YORK

Other books by LES LINE

PUFFIN ISLAND
MYSTERY OF THE EVERGLADES
WHAT WE SAVE NOW (EDITOR)
THE SEA HAS WINGS
SEASONS
DINING ON A SUNBEAM

A George Hornby Book

*The material in this book appeared originally,
in somewhat different form, in* AUDUBON *magazine.*

Library of Congress Catalog Card Number: 74–19002

Printed in the United States of America

Published simultaneously in Canada by General Publishing Company Limited

DESIGNED BY WALTER MILES

Composition by Hallmark Typographers, Inc.
Color Plates by Chanticleer Co., Inc.
Printing by The Case-Hoyt Corporation

Uncredited photographs: Forest floor by Eliot Porter; blue-eyed grass by Les Line; western meadowlark by Jon Farrar; white-tailed deer fawns by George Laycock; great egret by M. Philip Kahl; wild currant leaves by Bill Ratcliffe

For Michael and Heather
in the hope that their Earth
will be as good as mine

Contents

Introduction

AUDUBON. A man, a movement, a society, a magazine, a cause. The man, of course, is John James Audubon—naturalist, explorer, and artist of the early nineteenth century. Of his monumental work, THE BIRDS OF AMERICA, it has been written, "No other artist of his time was so sensitively and intimately concerned with the natural wonders of the American scene. No single person before or since has contributed so much to the art of painting birds." So it was predestined that Audubon's name would be adopted, half a century after his death, by America's fledgling bird protection movement.

Before the founders of that movement loomed an awesome task. A continent's treasure of birdlife, large and small, was being slaughtered at every hand—slaughtered for fun, for food, for fashion, for "science," slaughtered on migration in spring and fall and slaughtered on nesting grounds. Markets in large cities were crammed with the bags of commercial hunters—shorebirds, waterfowl, even songbirds. (In the spring of 1897, a single market stall in the nation's capital peddled the carcasses of 2,600 robins that had been shot in North Carolina.) Egg collectors, whose despicable hobby knew no morality, pursued every species, especially the rarest, to their nests. (One oologist, as they called themselves in assuming the respect of science, stole 917 eggs from 210 nests of the Kentucky warbler.)

Most heinous of all was the massacre of five million birds every year to adorn the hats and dresses of fashionable women. On two strolls down Manhattan thoroughfares in 1886, ornithologist Frank M. Chapman counted 542 hats that were decorated with mounted birds of 40 different species, among them warblers, woodpeckers, owls, and bluebirds. Mostly, however, it was the plumes of the wading birds, the egrets and herons of southern swamplands, that were prized by the millinery trade. The way that these feathers were obtained was vividly described by T. Gilbert Pearson, who one day would lead the Audubon movement:

"In the tall bushes, growing in a secluded pond in a swamp, a small colony of herons had their nesting home. I accompanied a squirrel hunter one day to the spot, and the scene which met our eyes was not a pleasant one. I had expected to see some of the beautiful herons about their nests, or standing on the trees nearby, but not a living one could be found, while here and there in the mud lay the lifeless forms of eight of the birds. They had been shot down and the skin bearing the plumes stripped from their backs. Flies were busily at work, and they swarmed up with hideous buzzings as we approached each spot where a victim lay.

"This was not the worst; in four of the nests young orphan birds could be seen who were clamoring piteously for food which their dead parents could never again bring to them. A little one was discovered lying with its head and neck hanging out of the nest, happily now past suffering. On higher ground the embers of a fire gave evidence of the plume hunters' camp."

Thus it was "to discourage the buying and wearing, for ornamental purposes, of the feathers of any wild birds, and to otherwise further the protection of native birds," that the first state Audubon society was organized in Boston, Massachusetts, in the winter of 1896. Its nucleus: conscionable ladies whose names were culled from THE BOSTON BLUE BOOK, plus field ornithologists and concerned sportsmen. It was an idea whose time had come. By the end of the year the Massachusetts Audubon Society had 1,284 members. A year later there were Audubon societies at work in Maine, New Hampshire, Connecticut, Rhode Island, New York, New Jersey, Pennsylvania, Illinois, Wisconsin, and the District of Columbia.

Now the Audubon movement needed a national voice, and so Frank M. Chapman, in 1899, launched the magazine *Bird-Lore*. The movement likewise needed national unity, and so in 1905 the then thirty-five state Audubon societies formed the National Association of Audubon Societies, whose goal was bluntly defined by the first president, William Dutcher: "The object of this organization is to be a barrier between wild birds and animals and a very large unthinking class, and a smaller but more harmful class of selfish people. The unthinking, or, in plain English, the ignorant class, we hope to reach through educational channels, while the selfish people we shall control through the enforcement of wise laws, reservations or bird refuges, and the warden system."

And for the emblem of the Audubon movement, the choice was obvious: a white egret on the wing against a blue sky. That image was a perfect rebuttal to the United States senator from Missouri who asked: "I really honestly want to know why there should be any sympathy or sentiment about a long-legged, long-necked bird that lives in swamps, and eats tadpoles and fish and crawfish and things of that kind; why we should worry ourselves into a frenzy because some lady adorns her hat with one of its feathers, which appears to be the only use it has?"

The protection of America's birdlife—through education and laws and sanctuaries—has, of course, been achieved, although federal sanctions against the killing of the misunderstood birds of prey were only recently obtained, and the survival of many species is now clouded by the insidious poisoning of Earth's air and water and land with deadly chemicals. The federation of state organizations that was founded seven decades ago is now one of America's most respected and influential conservation organizations, the National Audubon Society, 300,000 members strong. And Frank Chapman's modest journal has evolved into the world's leading voice for conservation, *Audubon*, from whose pages the chapters of THIS GOOD EARTH were drawn.

I recall the past with pride, in the belief that those readers who turn these pages should know something of the history of the movement that gave birth to the magazine that gave birth to this book. But also as an opportunity to dispel a myth of long standing (and suffering): that the National Audubon Society, and *Audubon*, are in the strictest sense for the birds.

It is quite true that for decades the Audubon movement was predominantly—but never exclusively—concerned with bird protection. It is also true that its magazine, until more recent years, was written mainly by and for bird-lovers, although many pioneering essays espousing a broad environmental awareness appeared in its pages.

But, as a recent letter to *Audubon's* editors pointed out, there are only two settings where birds do not interact with and depend on other species—in zoos and museums. The point he was making is that a person cannot truly appreciate the wonders of birdlife without an understanding of all of Earth's life processes, that one cannot truly be concerned with bird protection without an equal concern for the future of the whole environment. Thus birds no longer— indeed, cannot—dominate the issues of *Audubon*, as is quite apparent from the diverse contents of THIS GOOD EARTH. And while the National Audubon Society remains the leading bird conservation organization in America, this is but one of the many priorities of the modern Audubon cause, goals that range from conservation of all wildlife and its natural environment to pollution prevention to proper land use to a sane energy policy to the stabilization of human population.

Today's National Audubon Society member is interested in birds *and* bighorns and lichens and wildflowers and tortoises and sea urchins and prairies and mountains—in all things animate and inanimate that, combined, make up this good Earth. It is to refresh that appreciation—and to help bring an awareness of the infinite variety and fantastic complexity of our planet to the millions of people in America and England and the rest of the world who do not belong to an Audubon society or a sister conservation organization—that THIS GOOD EARTH is published.

The stories therein are neither tracts nor sermons nor predictions of gloom and doom. The pictures do not repeat the blight and destruction that can be witnessed at almost every hand. They are, words and images, a celebration of beauty, of goodness. We hope they will encourage whatever steps are still necessary, as they were three-quarters of a century ago, to keep Earth good, to protect it from the corruptions of the ignorant and the selfish.

One other matter: The publishing of a magazine demands a team effort, and there are members of the talented *Audubon* team who must share credit for this book. Ann Guilfoyle, senior editor, is responsible for tracking down the photographs that illuminate *Audubon's* pages and for shepherding them through the perils of full-color reproduction. Walter Miles, art director, quite naturally devoted his talents to the design of THIS GOOD EARTH. Kathleen Fitzpatrick, production editor, is the guardian of accuracy, style, and proper use of the written word. Roxanna Sayre, associate editor, presides over the magazine's corps of environmental reporters.

The stimulus for this volume came from Crown Publishing's creative wizard, George Hornby, who insisted that the book could be produced in an impossibly short time, and then stood by patiently while selection of stories and pictures and their layout waited for more pressing regular deadlines to be disposed of. As for the beauty of the printed page, credit is due in no small part to *Audubon's* lithographers for nearly a decade, the Case-Hoyt Corporation of Rochester, New York, and especially to Bill Lodgek who stands watch each time the presses roll. Over the same span of years, the color separations, that critical link between picture and paper, have come through the skilled hands of the Chanticleer Company of New York City.

LES LINE
Editor, *Audubon*

This Good Earth

One day, 50 million years before man, a leaf fell from a sycamore into a stream in the tropics of what is now called Utah. Carried to a still place in a vast inland lake, the leaf sank and was covered by sediments. But unlike leaves that turn to humus on a forest floor, it left an imprint so perfect that it could be identified eons later.

The Record in the Rocks

PHOTOGRAPHS BY BILL RATCLIFFE

TEXT BY RUTH MOORE

THE ROAD climbing the precipitous western range of the Andes had turned into a narrow zigzag track. The mules carrying Charles Darwin on his first crossing of the Cordillera in March of 1835 halted every fifty yards for a rest, and Darwin surveyed the "glorious" scene—the broken jagged pinnacles, the profound valleys, the condors wheeling silently in the intense blue of the sky—but his eyes also were caught by a seashell. He saw a fossil shell and then numberless shells protruding from a broad band of pale limestone along the rock face.

Darwin knew that he was at an elevation of about 13,000 feet, close to the summit of the mighty mountains. Fossil shells at 13,000 feet! His muleteers had been complaining about the *puna*, the shortness of breath in that rarefied atmosphere, and Darwin was experiencing a tightness across his head and chest. Nevertheless he quickly dismounted and began to collect the shells. He pried out a *Gryphaea*, an *Ostrea*, a *Turratella*, some ammonites and small bivalves. He would have "reaped a grand harvest" if he had not been forced to hurry. But it was late in the season and if snow began to fall their situation could be dangerous. He also could miss the sailing of the *Beagle*, the ten-gun brig on which he was circumnavigating the Earth.

As Darwin gathered his fossil shells he saw immediately that some of them were similar to the shells he had collected on the beaches far below. He recognized that these shells at one time had rested on that ocean bottom. Through untold workings of Nature, the onetime low-lying beds had been uplifted to the nearly unimaginable height of 13,000 feet.

On the eastern slopes of the mountains Darwin found the petrified stumps of trees that once had also been under water, and more fossils. The mountains in part were made up of sedimentary rocks. The Andes, then, had not been created wholly by the outpourings of volcanoes, as the prevailing theory held. There had to have been uplift.

"I know some of the facts, of the truth of which I in my own mind feel fully convinced, will appear to you quite absurd and incredible," Darwin wrote to a scientist friend.

*A butterfly, buried 35 million years ago
beneath volcanic ash at Florissant, Colorado . . .
And from Utah's Green River shale formation,
two 60-million-year-old leaves, one of a poplar,
one from a* Caesalpinia, *a vast tropical genus
of trees and shrubs that includes Brazilwood.*

The fossils Darwin collected on the crossing—half a mule's load—had shaped a new theory of the origin of the Andes. The fossils also were helping to change Darwin's and ultimately the world's concept of time. For the shells to have been uplifted from the sea floor to crest would have taken millions and millions of years, even though these mountains were young in comparison to some in other parts of the world.

The fossils buried in the rocks were in one sense time clocks. They also spoke of the order of the past. As long as there was no disturbance, the oldest strata and the oldest shells lay at the bottom of any series. The younger sediments, and the younger fossils, had been deposited on top of the older. If the strata were in their natural order, the order was as clearly visible as that of a stack of books.

The fossils, furthermore, were the actual record of the past. Year after year and millennium after millennium the remains of countless numbers of sea and land creatures drifted to the bottoms. Most were destroyed, but many hard parts, their prints, and their casts were preserved as they were covered by the sands and silts of the floors or by the continuous rain of other bodies and materials settling down from above. On dry land only a few were preserved in asphalt pits, bogs, or the ice of the Arctic.

Long before Darwin discovered the fossil shells high in the Andes, men had been finding fossil shells and bones weathering out of the earth. Some of the Greeks had picked up seashells well inland and had reasoned that the sea must once have rolled over the land where they lay. Occasionally a huge piece that looked like the bone of an animal came to light, and was attributed by the Greeks to some kind of monster.

During the next 2,000 years fossils continued to turn up. Few, though, accepted the Greek explanation, if they knew of it at all. Learned men argued that the oddities had grown in the earth from seeds fallen from the stars. Others insisted that they had been formed in the ground in chance imitation of life. Still others maintained that they were the works of Satan, put into the ground with the deliberate intent of deceiving overly curious men.

By the latter half of the seventeenth century fossils posed a new kind of problem. Scholars had discovered with a shock that the Bible might be subject to various readings and translations. To

*Lovely and delicate in death, and a record in the rocks of a time
on prehistoric Earth that modern man can scarcely conceive,
are these fossil leaves of an ancient sumac, Rhus stellariaefolia.*

uphold the Biblical authority upon which society and religion rested, religious leaders undertook to prove by the science of the century that the miracles and particularly the account of creation were true. Scientific certainty was to support revelation.

Fossils thus had to be accounted for. John Ray (1627–1705), a Cambridge University clergyman, was too good a naturalist to attribute fossils to chance or the devil. He saw that some of the fossil shells he collected were exactly like those washing up on the beaches. Others, however, were the skeletons of fish that he knew lived only in the deep oceans. He was also troubled by the *Cornua ammonis* or serpent stone. He had to classify it as a species unlike any existing one.

Ray did not question that the fossils had settled into their resting places during Noah's flood. But why were they not spread evenly across the Earth, instead of amassed in "the great Lumps" in which he found them? And how could an unknown species exist, if all species had descended from those Noah rescued in the ark?

Ray concluded that the waters of the abyss—reservoirs in the bowels of the Earth—and all the surface waters with which they were connected were filled to overflowing during the forty days and nights of Noah's downpour. Under the tremendous pressures, the floodwaters burst forth "at those wide Mouths and Apertures made by the Divine Power breaking up the Fountains of the Deep." The waters engulfed the Earth, carrying with them the fish and other life of the sea. Perhaps some of the fish and shells washed up even into the highest mountains.

John Woodward (1665–1728), another professor-clergyman who was struggling with the problem of his day, had a different theory of the flood, but agreed that fossils were "all remains of the Universal Deluge, when the Water of the Ocean being boisterously turn'd out upon the Earth, bore along with it Fishes of all sorts, Shells, and like moveable Bodies."

Fossils and their occurrence thus were satisfactorily explained. All was once more in order.

Less than a century later, however, new fossil finds put the whole problem into new terms. Georges Cuvier began to find the fossilized bones of elephants, flying lizards, and other fantastic animals in the soil of the Paris area itself. Fashionable Paris, all agog, rushed out to the gypsum quarries where he was unearthing them to see the spectacle. Cuvier increased the sensation when he restored many of the animals to the appearance they had in life.

"Is Cuvier not the greatest poet of our century," exclaimed Balzac. "Our immortal naturalist has reconstructed worlds from blanched bones. He picks up a piece of gypsum and says to us 'See!' Suddenly stone turns into animals, the dead come to life, and another world unrolls before our eyes."

The strange menageries of the past that Cuvier was producing, almost sorcererlike, were not a random assortment. The animals of the past, like those of the present, belonged to species and orders. Furthermore, most of the orders were represented—birds, mammals, reptiles, and the others.

Cuvier and Alexandre Brongniart, the famous Sèvres porcelain maker and geologist who worked with him, also saw that the sea fossils lay in one stratum, the land fossils in another, and

18

then that there might be a layer with no fossils at all. At one time, they understood, the seas had covered the Paris Basin, as the area was called. Again the waters had receded, and the fossils of land animals were found in deposits laid down in lakes and rivers. Many such alternations had occurred.

The two scientists pushed their investigations farther into the country. They did not always find the same succession of strata, but they could identify those they found by the fossils they contained. If a bed of marl, a crumbly mixture of clay and limey materials, had the same fossils as the marl in the Paris area, they knew that it had been laid down at the same time.

"It is a method of recognition that up to the present has never deceived us," they wrote.

They remarked too: "The characteristic fossils of one bed become less abundant in the bed above, and disappear altogether in the others, or are gradually replaced by new fossils which had not previously appeared."

At almost the same time, the same discovery was being made in England. William Smith, a surveyor and engineer engaged in laying out canals, saw that "every structure contained organized fossils peculiar to itself," and might be "recognized and discriminated" by those fossils. Smith prepared a great map of England—a classic of cartography—showing "the same species of fossils are found in the same strata, even at a wide distance."

The observations—Cuvier's and Smith's—were new. Though most scientists of the time were willing by then to grant that fossils were the remains of actual animals, few if any had suspected that they were a clear record of the past and that they could be used to determine the order and organization of the past.

Fossils helped, perhaps determinatively, in building Darwin's theory of the evolution of life. On the voyage of the *Beagle* Darwin also had collected the fossilized bones of nine ancient animals. The *Toxodon* had been as large as an elephant, but it had the teeth of a gnawer.

"How wonderfully are the different orders, at the present time so well separated, blended together in different points in the structure of the *Toxodon*," Darwin wrote in his travel journal.

Later, as he pored over his fossil specimens, Darwin was forced to admit to himself that the fossil animals must have been the ancient ancestors of the living. That could only mean that the living species had not been separately, miraculously created. This revolutionary insight was a major one leading Darwin into the twenty years of work that produced his theory of evolution. All living things, he maintained, are descended from earlier forms and ultimately from the first living cell.

If this were true, his angry and shocked critics demanded, where was the proof? Where were the missing links, and most particularly the links that would join man to apelike ancestors?

Darwin could only answer that the Earth had not been widely explored for whatever fossils it might hold. Though many opponents were certain there was no record to be discovered, the search was on. Expeditions were sent out; individual scientists, and many amateurs, were on the

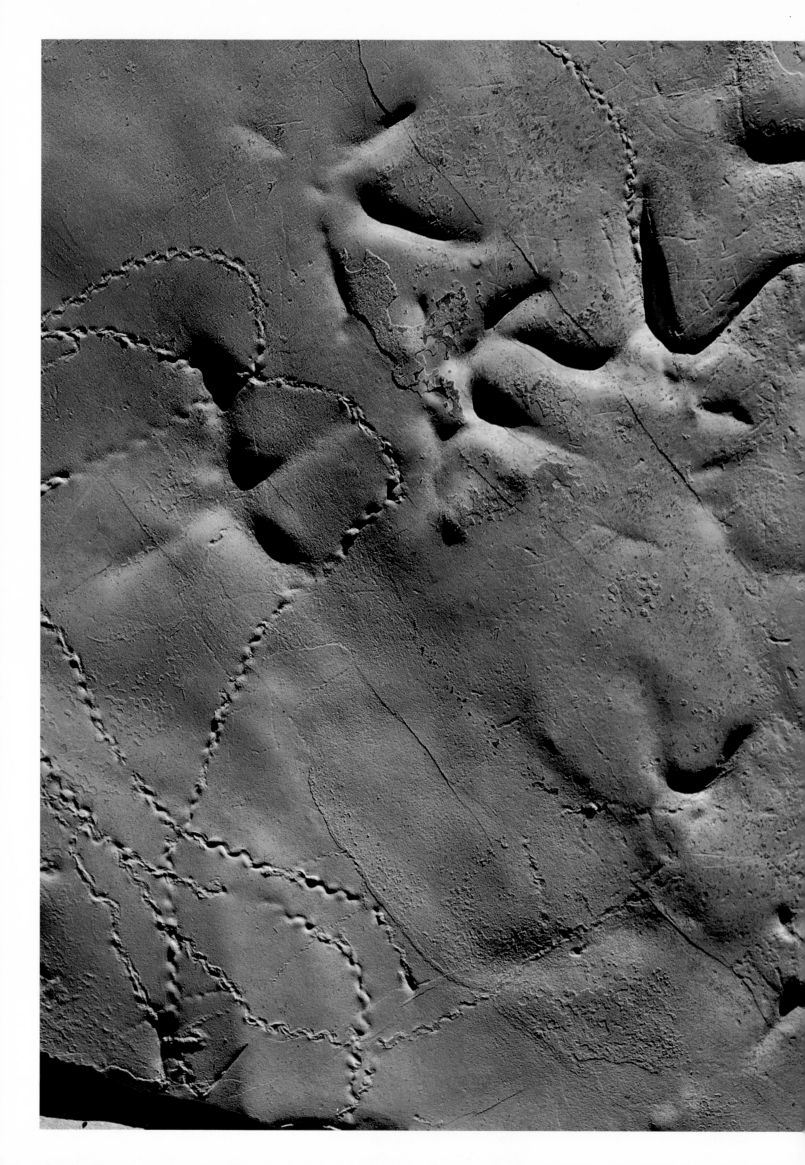

Birds, large and small, and insects unknown walked the shoreline of ancient Lake Uinta
in Utah during the Eocene Period of 60 million years ago. Some birds were the size
of sandpipers. Some were web-footed like ducks. Some left heronlike tracks in the mud.
And those impressions were protected with sand or more mud as the rock-making process
went on. One hundred million years earlier, not too many miles away, a three-toed, 20-foot-long,
meat-eating Allosaurus left its footprints in future sandstone near the Colorado River.

lookout for fossils. It was less a matter of digging than of finding the fossils as they began to weather out and then of extricating them. Only a minute part of the life of the past had been preserved, but that was significant. Find followed find.

In a quarry in Bavaria, Germany, workmen quarrying the extremely fine lithographic limestone came upon a remarkable fossil cast and imprint. Nothing like it had ever been seen before. It had a long head, sharp teeth, an extended neck, and strong hind legs—all reptilian characteristics. But the fine impress also showed in exquisite detail that feathers had clothed the body. If the feathers had not undeniably been there, no one probably would have imagined their presence. Was it bird or reptile? Later two other casts were found, although they were not as vivid as the first. Seemingly the bird–reptiles had fallen into the waters of the clear coral lagoon that once occupied the area, and as more fine lime settled down over the bodies they were preserved, with some of the parts shown in as true and sharp detail as the greatest sculptor might achieve in a cast for a bronze medallion. Later, the soft ooze that served as the mold hardened and the cast was preserved indefinitely.

With the stone to testify, the feathers had to be adjudged true bird feathers, and the creature was named *Archaeopteryx*—one of the earliest and most primitive of a group that in time would take over the province of air.

Not until 1970 was a fourth *Archaeopteryx* found. It was discovered not in the field but in the Teylers Museum in the Netherlands. It was on display, but incorrectly identified. The museum records showed only that the slabs of limestone bearing the impress had come to the museum before 1857. The newly found bird was less spectacularly engraved than the first, now in the British Museum (Natural History), but the long, crescent-shaped claw on its wing had survived uncrushed. This claw, appearing so unexpectedly out of the past, was expected to help in determining whether these early birds were arboreal- or cursorial-runners.

Such smooth-grained limey lagoon floors and other fine muddy and sandy shallows also yielded the crawl-trails and often the footprints of early crawlers and waders. In beds of stone now raised high and dry, the fossil hunters not infrequently came upon the wandering cord or scratchy path made by some insect millions of years before. The trail of a tiny crab had won a permanence far exceeding that of any work of man. The footprints of birds frequently were all about in such stone.

Other tracks and footprints outlined the feet and the courses of some of the largest animals. Before one set of giant three-pronged footprints was identified as those of a *Brachiosaurus*, a dinosaur that sometimes grew to eighty feet in length, Mark Twain wrote an article about them. He attributed them to the imaginative joking of some imbibers in a nearby tavern.

At places the fossil search disclosed that miles of thick beds of limestone were compacted of the shells of inhabitants of the seas. Many of the shells, like those Darwin gathered high in the Andes, were preserved intact. Their hard structure had lasted limitlessly. In others, chemical

24

change had occurred. Occasionally the original structure of shell, bone, and wood had been replaced by minerals dissolved in the waters. It had been petrified in the true meaning of the term, though such transformation was more rare than common. The sea floors and the water-laid rocks of the continents were in fact vast graveyards of the past.

How immensely long and complete some of this compiled record could be was discovered only in the 1950s and 1960s when the Lamont-Doherty Geological Observatory at Columbia University and, more recently, the Deep Sea Drilling Project supported by the National Science Foundation, recovered long cores from the ocean sediments and underlying floors. The drills of the special exploratory ship the *Glomar Challenger* were dropped through as much as three miles of water to cut cores through more than half a mile of sediments on the ocean floor. The long tubular cores with layer after layer of fossil and other deposits pictured the history of the Earth through periods of tropical warmth and glacial cold, through the constant movements of sea bottoms and through the evolution of life. For long periods the record was continuous. On land it was seldom so, for there had been many times of erosion when no strata were accumulated.

Even in the deep oceans, no one core bore the entire record. Deposition had differed in various areas and in the different oceans. But by studying the many cores, the thousands of cores cut in the two decades, scientists began to bring together the whole story.

Unexpectedly, however, the fossils and the layered cores helped essentially in providing new insights into the form of the Earth's surface, the origins of its mountains, its volcanoes, its earthquakes, and its tidal waves.

With their soundings and seismograms, Columbia University scientists first discovered continuity of the 47,000-mile chain of mountains running approximately midway through the Earth's oceans. Some of the underwater mountains were higher than Everest. Here, it was soon determined, lavas were constantly welling up from below. New additions were steadily being added to the ocean floor. When cores were cut in these areas of replenishment, the sediments and the animal remains were modern. In geological terms, they were new. Sailing 150 miles outward from the ridge, other cores were taken. Here the fossils showed an age of about 10,000,000 years. At 300 miles away from the ridge the age was 20,000,000 years, and at 750 miles distant, 70,000,000. It was the same on the other side of the ridge. Age increased as the ships moved to either side of the ridge. Measurements of radioactivity and of the magnetic alignment of the mineral grains in the rocks—fossil magnets—confirmed the same increase in age as distance increased from the ridges.

A startling conclusion was reached. The sea floors were being pushed outward as new material pushed up from the lower parts of the crust. In some cases the movement amounted to several inches a year. In a few areas, the scientists calculated, a sea floor might move as much as a man's height in his lifetime.

As the push continued through millions of years the abutting continents were shoved along, too. When a westward-moving block, like that made up of the western half of the Atlantic floor and the American continents, met an eastward-moving block, such as the eastern half of the Pacific floor, the result was collision, the greatest of all collisions. As in many head-on clashes, the lighter body, here the continents, ran up and over the heavier Pacific floor. The continental edges were upthrust into the long and towering coastal ranges of both North and South America.

At the same time the sixty-mile-thick Pacific floor was crunched down and under, and back into the lower depths of the crust. At the point of downturning were the oceanic trenches that lay just offshore.

The conflict and pressures generated in this colossal meeting of forces produced the earthquakes, the tidal waves, and the volcanoes that historically and prehistorically had rocked the Pacific rim.

For the first time an understanding was being reached of the previously unexplainable continents and seas, of the Earth's mountains, and its upheavals.

The fossil cores also helped to demonstrate that all the continents may at one time have formed a single land mass. As upwelling along ridge areas started outward movements, the one mass was broken apart. North and South America moved from their original position next to Africa and Europe to their present positions. And as the New World continents moved westward the Atlantic Ocean filled in the rift that was created. The findings were revolutionary. In a few years they began to remake the concept of the Earth. The change was compared to the Darwinian revolution and the atomic breakthrough in physics. Without the fossil cores the understanding might have been much slower in coming.

On the continents—now recognized as the oldest part of the Earth in contrast to the frequently remade and thus younger sea bottoms—the fossil search went on apace. Long before the movement of continents was confirmed or imagined, science knew that there had been vertical movements in the Earth's surface. Wide shallow seas had repeatedly washed over the whole central basin of North America and over the lower parts of other continents. As in the seas, the bones and shells of land and sea animals had drifted down to the bottom of these inland waters. The skeletons of other animals were washed down into rivers and lakes. Only in the rarest of cases did the flesh and the soft parts survive predators and decay.

A fevered part of the early search for American fossils centered in the West. Two rival fossil hunters fought for priority and for fossils.

Shortly after the Civil War, Othniel Charles Marsh, professor of paleontology at Yale University, found huge fossil bones in western Kansas. This was a part of the continent that had been covered many times by the advancing and retreating inland seas. After the last retreat of the water the land had risen and erosion had set in. It exposed many of the ancient seabeds and their holdings of the past.

Pioneers making their way west often saw some of these fossils, and generally assuming that they represented the backwash of Noah's flood, went on by. The area was a rich and nearly untouched treasury of fossils when Marsh arrived.

Edward Drinker Cope, a young Philadelphia Quaker who had decided to devote his life and not inconsiderable fortune to the search for fossils, was drawn to the same promising field. Marsh, however, considered the territory exclusively his own.

Cope had only arrived when he saw the bones of a fossil reptile projecting from a bank above a dry creek. When he and his men dug it out, it proved to be nearly 75 feet long. That night, as he sat close to his campfire and described the monster, Cope named it *Liodon dyspelor Cope*. It was the first of more than 23,000 specimens, and 463 species, that Cope was to collect and name with similarly formidable titles.

Marsh collected as widely; both men often wired the names of their discoveries back East to assure priority. It was one of the classic rivalries of American science. But out of the competition and urgent search came the denizens of several millions of years.

The fossils collected throughout the world were showing finally how life had advanced from fish to the amphibians, to the reptiles, the birds and mammals.

For many years the records in the rocks stopped short of man. Cuvier had argued with his monumental authority that there was no such thing as fossil man. Darwin could cite no fossils between apes and men. At the very time, though, that he was being pressed by his critics, an unusually thick, low-browed skull had been found in the Neanderthal Valley in Germany. Some authorities insisted that it was only a pathological specimen. Darwin's friend and supporter, Thomas H. Huxley, warned that "Neanderthal man" could not be regarded as an intermediate —he was a man, though an unusual one. But Huxley asked if in some older strata some future paleontologist would not find the fossilized bones of an ape more anthropoid (manlike) or of a man more pithecoid (apelike) than any yet known.

In 1887 Eugène Dubois, a young Dutch physician, sailed for the Dutch East Indies (now Indonesia) to search for this creature. Improbably, he found him. *Pithecanthropus erectus*, as Dubois named him, was an early man with a low skull but with a thigh bone that indicated that he had walked upright, like a human. Up to this point the scientific world had expected that if any predecessors of man ever were found they would have exactly the opposite characteristics— a large brain and the stooped, shuffling body of an ape.

Outrage was extreme. Clergymen proclaimed that Adam and not this crude freak from Java was the ancestor of man. The anger had not abated when a somewhat similar fossil skull was found in a filled-in cave at Choukoutien, near Peking.

The controversy was still raging when, in 1927, Dr. Raymond Dart, collecting in South Africa, found the fossil skull of a six-year-old child with a brain case no larger than that of a similarly aged ape, but with other clearly human traits. Not until many years later when many adult Australopithecine skulls—Broom's name for the group—had been recovered in South Africa, and the work in the Orient had yielded many leg bones, feet, and pelvic bones to attest to the upright and human structure, was it acknowledged that these were among man's ancestors. This was the way it had come about—ape-size brain and human body. The evidence of the bones, the fossils, was no longer to be denied.

Thus the record was filled in, even up to man. Life could be traced in the rocks for nearly half a billion years. The fossils evidenced history and time.

In the famous fossil beds at Florissant, Colorado— now a national monument—stands a giant petrified stump of a redwood that lived 5,000 years and stood 350 feet high.

Lava Is Life

PHOTOGRAPHS BY DAVID MUENCH

TEXT BY DAVID LEVESON

To FAMILIARIZE the astronauts in advance with the stark, lifeless character of the moon, their training includes a sojourn in the awesome volcanic landscape of Hawaii. There, desolate arrays of newly formed cones and craters alternate with glistening lava fields. Much of the moon's surface, except where covered with loose debris or modified by meteoric impact, is similar. Basaltic lava is the basic ingredient: dark, solidified molten rock like that commonly erupted from terrestrial volcanoes.

For those of us who live in North America (with the exception of the Aleutian Islands, Mexico, and the islands of the Caribbean), volcanism and molten lava are exotic phenomena, something to be found across an ocean, in Hawaii, in Italy, or, indeed, on the moon—but not here. Aside from the eruption of Mount Lassen in 1914–1921, volcanic activity in the contiguous forty-eight states and Canada has been in abeyance for almost the last one hundred years. The only reminders of the Earth's violent, fiery, subterranean temper are a few dozen geysers, associated hot springs, some rare pools of boiling mud, and those places where evil-smelling gases whistle out of the ground.

But this period of calm is likely to be brief, as is soon suggested by a glance at a geologic map —a map that shows the types and ages of the rocks that underlie the ground we tread so confidently. The evidence is clear. Within Oregon, Washington, and Idaho, tens of thousands of square miles have but recently solidified and cooled from molten rock extruded to the surface in great white-hot sheets. In Nevada, Utah, New Mexico, and California, areas of equal magnitude are underlain by volcanic ash—formed from lava that exploded as it reached the surface and poured over the land as an incandescent mixture of gas and solids, or filtered down as a suffocating, unthinkably hot dust.

Continued on page 37

30

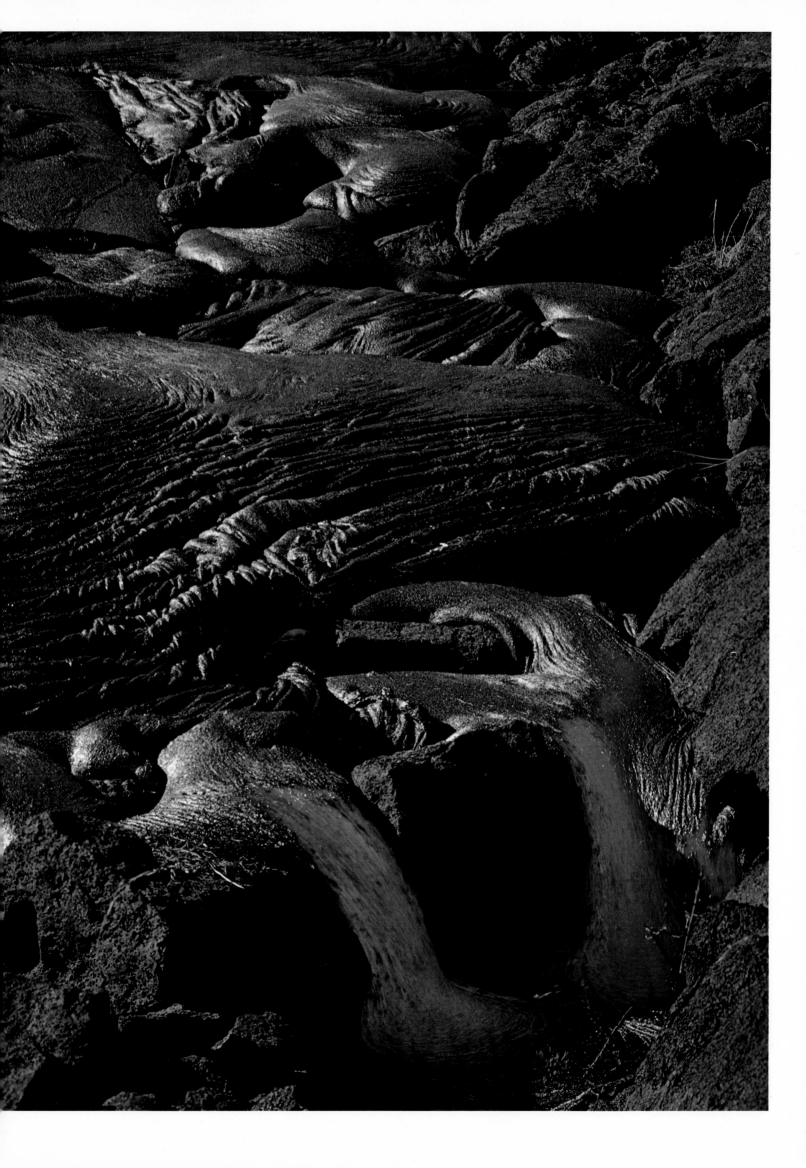

Rabbitbush and sunburnt clumps of Russian thistle, one of the tumbleweeds,
decorate the cinder-strewn flanks of Merriam crater in Arizona's San Francisco
volcanic field. In bloom the rabbitbush is covered by tiny yellow flowers
which the Navajos use to prepare medicines for coughs, fevers, colds.
And monkeyflowers brighten the cindery rubble slopes of the Great Rift
at Idaho's Craters of the Moon. In the distance, a cluster of cinder cones
forms an irregular horizon. Less than 2,100 years ago, this peaceful scene
saw a violent, fiery eruption and a hail of incandescent lava fragments.

33

Silversword stands in stark relief against the cinder cones and craters of Maui's
Haleakala National Park. And at Hawaii Volcanoes National Park, the underlying lava
is obscured by a rich garden of lichens, club moss, ohia lehua, and amaumau fern.
Cinder cones hundreds of feet high, with bowl-shaped depressions at their tops,
may be created in a matter of days from the porous bits of solidified lava
—usually less than an inch in diameter—that are expelled during violent eruptions.
Oxidized and hydated iron then stains the cinders red, orange, and brown.
But gradually air, water, and living organisms turn the volcanic rock into fertile soil.

Why aren't we more generally aware of this? Destruction and waste have been transmuted. The harsh contours of frozen lava have softened, the sterility of dry ash has been quickened. With time, rain and air and the persistent, blindly optimistic activity of primitive organisms have turned rock into soil, and the soil has become colored with life. How much time? A long time, measured in human terms. Thousands, millions, even tens of millions of years may have elapsed since the roar of volcanic eruption ceased. But when measured against the four-and-a-half-billion-year length of *geologic time*, it seems as if the Earth has hardly ceased to tremble.

In a few places, extensive volcanic activity was recent enough to be recorded in myth and tribal memory, and the ground is still an almost lifeless desert, where liquid shapes remain sharply poised, seemingly ready and able to advance again. In Idaho, the dramatic, alien nature of volcanic landscape has been formally recognized as Craters of the Moon National Monument. Fantasy successfully anticipated the experience of future astronauts by several decades. Gaping craters, hardened ropy lava, and angular blocks of cinder form a mordant landscape essentially unrelieved by life. The fragility of human existence in a universe of inhospitable extremes is inescapably apparent.

But if lava suggests lifelessness, it also suggests life. Each atom of the air we breathe, the water we drink, stems ultimately from vapors that have bubbled up to the Earth's surface through the narrow throats of volcanoes. Originally airless and waterless, the Earth has achieved its atmospheric and oceanic mantles through *geologically* gradual accumulation of volcanic gases released by erupting lava. Thus lava is life, pregnant with potential, beautiful in its diverse forms, awesome in its destructive capability, a constant reminder of mortality.

At Idaho's Craters of the Moon National Monument, an ancient river
of lava is frozen in time. Lava may flow at speeds up to 25 miles per hour
and resemble a fiery stream. But most fluid lava is 100,000 times less fluid
than water. And as the lava cools and the gases contained within it
are released, mobility decreases rapidly and it congeals into a solid.

Of Wind and Coral Sand

PHOTOGRAPHS BY BILL RATCLIFFE

TEXT BY JOSEPH WOOD KRUTCH

EVERY YEAR thousands of tourists pass through the little Utah town of Kanab, heading northward toward Zion and Bryce canyons or eastward toward the new Glen Canyon National Recreation Area. Most of them do not even notice a modest wooden arrow a few miles east of the town which points down a dirt road and is marked "Coral Dunes." Still fewer accept its invitation, and that is perhaps just as well, for the undisturbed beauty of the dunes (some dozen miles from the main road) is one of their charms. On an early June morning, my companions and I had them to ourselves in the cool of five thousand feet.

Much of northern Arizona and southern Utah is a land of towering sandstone mesas and buttes, some white, some pink, and some coral red. They have been sculptured into fantastic shapes and are gradually being eroded away by water, frost, and especially by wind-blown sand. Some are half-buried in their own detritus but still rise sheer above the semidesert plains. At the Coral Dunes, on the other hand, the prevailing wind has heaped and shaped sand into dunes as high as twenty-five feet, which, incidentally, recently furnished a perfect setting for a motion picture, whose action is supposed to take place in the Arabian Desert.

Nothing quite prepares one for the climactic view. The approach is across a semidesert, increasingly sandy but with the sand held in place by fairly abundant sage and juniper together with a few pines. All of them grow less and less abundant, and then one comes upon a true Sahara of drifting coral-colored dunes sloping gently upward on one side, dropping off abruptly on the other; sometimes rippled as though by waves of a seashore; sometimes almost unbelievably smooth and sleek.

At first, one is unaware of any living thing except, perhaps, for ravens calling derisively overhead. No animals are visible. But the most casual inspection reveals the fact that they are only unseen, not absent. There are tracks which can only be those of a bobcat and there are other

38

To the geologist, the dunes tell one story—
of a building up and tearing down and
rebuilding. To the ecologist, they tell
another story—of a struggle to survive,
of a sunflower which has managed to set
its seed in the nick of time, for it will
soon be buried while, in the process,
changing in its own small way the face
and shape of the dunes. And to the esthete
the dunes speak incredible beauty and form,
the lanced shadows of yucca pointing to
desert puffballs, the fragment of tumbleweed
uncovered by the whims of the wind,
looking so much like the vertebrae
of some long-perished dunes denizen,
or merely the effect of late sun on sand
already colored coral by its heritage.

curious little bipedal marks which proclaim the kangaroo rat. Most beautiful, and at first most puzzling, are long lines of the most delicate tracery, sketched across the smooth surface uphill and down. Each one is perfect despite its obvious fragility and all suggest some secret mysterious workman. What could have made them? Not a small lizard because there is no trailing tail mark. Certainly not a sidewinder rattlesnake, whose strange tracks are broader and quite different.

It doesn't take much looking to find the answer. The tiny workmen are everywhere busy, often no more than a few yards from one another. They are little, shiny, quarter-inch, scarablike beetles (*Sphaeriontis muricata*). From the order to which they belong, it seems a pretty safe guess that their unresting progress uphill and down is motivated by nothing more spiritual than a search for the droppings of a jackrabbit. But it would not be difficult to imagine that they are artists, endlessly engaged in beautifying the dunes with the perfect but changeless pattern of lace which Nature has ordained their six legs to make during many millennia and which they will continue to make for untold millennia hence—unless, as seems not improbable, man destroys their environment as he is destroying that of many more conspicuous creatures.

Most of the dunes are shifting, and they are often marching forward in the direction of the prevailing wind. Any sizable plant which manages to get established tends to anchor the sand around it, but more often than not it loses the struggle and is overwhelmed by the slowly advancing waves which may ultimately pass over it, leaving behind a shallower bed of sand in which a new generation of plants may be able to grow.

Evidence of that process is plainly visible in the Coral Dunes. One may see, for example, the skeleton of a pine killed sometime in the past by an advancing wave; beside it are younger pines or junipers which have grown since the crest of the wave passed by. On the leeward side of one of the highest dunes a single clump of sagebush is visible. But it is half-buried and doomed.

Many wildflowers, able to establish themselves and mature in much less time than the juniper or even the sagebrush, flourish in patches which here and there provide an arresting flash of color, standing vividly against the red of the dunes. Two are especially conspicuous—the large yellow member of the sunflower tribe (*Wyethia scabra*) sometimes called "mules-ears" and the lovely blue *Sophora stemophylla* which, from a short distance, might be mistaken for a lupine and which has no common name. Most surprising in the desert is a little mushroom, the desert puffball (*Tulostoma poculatum*).

All the larger plants must send down deep roots, and some of them may survive even an advancing wave of sand by raising their stems higher and higher as the roots go deeper and deeper. This is the method of a handsome yucca of limited distribution named *Yucca kanabensis* after the nearby town. A different solution to the problem is to spread out over a large area and send down two shallow roots every foot or two to snatch from the surface the water of the rare and usually limited showers. This is the method of another characteristic plant, the scurf-pea (*Psoralea lanceolata*), of which a single prostrate branch may be more than seventy feet long.

Dunes like these are both esthetically pleasing and ecologically instructive. But they are equally interesting in another way. They occur less frequently than mountains or plains or valleys

Spring wildflowers push through sand like crocuses through lingering snow.
The sand dock—glowing red on a slope speckled by the few raindrops
of a brief shower—is the "wild rhubarb" relished by Indian tribes.

Wind. Plus sand. A dune marches and creates.
Here grains of sand—eroded sandstone and neither too large
nor too small—are rolled and barely lifted over a gentle slope,
to repose on the steep leeward side. And there, a ponderosa pine
has been smothered by a dune's advance, then etched by sandblasting.

On the following pages: What Omar called "snow on the desert's dusty face"
is no uncommon phenomenon. It may be a light snow, a windblown decoration.
Or it may be an icing deep enough to half-bury the dried tumbleweed.
Snowdrifts themselves are merely dunes made of ice crystals instead of
quartz crystals, and the same physical laws determine the forms of both.

or canyons, but like all of these they are quite distinct and recognizable geomorphic features and, again like the others, they raise in any inquiring mind the question of how and why they came into being.

In terms of earth and atmospheric mechanics, the answer is always the same whether the material be the sands of a seashore, the gypsum grains of the great White Sands of New Mexico, the eroded sandstone of these Coral Dunes, or even the ice crystals of the snowdrift. And the forms assumed are so similar that a photograph of the White Sands might easily be mistaken for that of a New England snowbank.

Given a flat, open surface and hard grains of material too heavy to be blown away by the prevailing wind but not too heavy to be moved by it, the dunes are nearly inevitable. If the grains are too light, they blow away as dust; if they are too heavy (small pebbles, for example), they cannot be piled up by the winds. But if they are just right, you get the similar outlines of the same geomorphic feature. Dryness helps, too, because damp sand doesn't blow easily. And there are minor variations in shapes, depending partly upon the character of the winds. But a dune is a dune is a dune.

Geographers tell us that the wind seldom lifts sand particles more than a foot or two, that the grains are usually simply rolled along, and that they come to rest when the wind's velocity decreases. Once started, a dune itself becomes an obstruction around which the winds swirl and deposit more sand. Because of the variations in wind velocity and sand supply there are different characteristic shapes in different regions, and those most characteristic of the Coral Dunes are what are called "barchans"—that is, hillocks with a sloping side up which the prevailing winds blow the sand and a steeper leeward side where the grains have come to their angle of repose.

Though the Coral Dunes of Utah are in many respects typical, there is one fact about them of additional interest: the material of which they are composed has been twice reduced to sand in the course of many millions of years.

The mesas and buttes which surround them are composed of Navajo sandstone formed during the Jurassic period (say 100 or 150 million years ago). But that sandstone was composed of the detritus from much more ancient mountains long before worn down. Now, this Navajo sandstone has itself been eroded away to make the sand which may (given more millions of years) again solidify into stone—either under water or, like the sandstones of nearby Zion, right on the desert itself, in which case they will still exhibit by their crossbedding the outlines of successive dunes, like those so clearly seen in Zion.

There are few more striking examples of the restlessness of our earth—always building up, tearing down, and then building up again as the result of the processes which will probably continue until a completely cooled earth can no longer raise mountains and its whole surface is reduced to a featureless plain.

There are no slums in nature, an ecologist once wrote. What he meant
is that every environment is a good one for the creatures which occupy it.
Dunes may seem to submit this statement to a severe test, and they may be
quickly dismissed as lifeless. But in truth they boast their quota
of birds, mammals, insects, reptiles. The delicate tracery across this
coral sand is the artwork of a small but abundant beetle that has threaded
its journey between the imprints of a kangaroo rat and those of a kit fox.

Lichens–Mirror to the Universe

PHOTOGRAPHS BY BILL RATCLIFFE

TEXT BY DANIEL McKINLEY

THIS ESSAY on small green things and their worlds was written while the wine of a country hike in late winter simmered in my veins. If there is in it a temporary sanity, it is due to the clarity of the sky, the evening sounds of birds, and the exuberant cleanness of the air. Lichens are the minutiae of the plant world. They are not invisible, yet they are not large enough to have lost the elfin qualities that go with perfection in small packages. And to diminutive size they add the charms of unexpected color and exquisite symmetry of form.

My interest in lichens is a hobby-interest. Nature hobbies—unprofitable and utterly fascinating interests in some small aspect of life—at worst are not more than harmless diversions. At best, they pull us out of our pride and enable us to see the world we live in.

The hobby world is a three-cornered hat. One corner is *identification*. The hobbyist who goes into the field to identify lichens will find a lifetime of interest and pleasure before him. These small green plants will take him from field books to his knees out in the field itself; and they may as easily transport him into the microworld by means of hand lens and microscope.

Another corner of the triangle is *life history*. Every fact gained is a piece of colored glass or broken stone that fits somehow, somewhere, into the puzzle that each kind of lichen is. "Cryptogams" they have been called, meaning "hidden marriage" and signifying the seeming complexity of their reproduction. While it is now well known that the lichens are composed of two entirely different kinds of plants (a fungus that houses some sort of food-producing, usually green and single-celled alga) living in close and apparently mutually beneficial union, it is still a tantalizing mystery why people in laboratories cannot mimic Nature and put them together successfully at will. Here, aside from complications that arise when "one" plant must give rise to offspring that are dual in nature, is a challenge. With such variations on the lichen theme extremely common, the unraveler of lichen life histories need never fear that the job will end and he will be pensioned off.

The hobby-hat's final angle is really the corner that I wear pointing forward. This is the

50

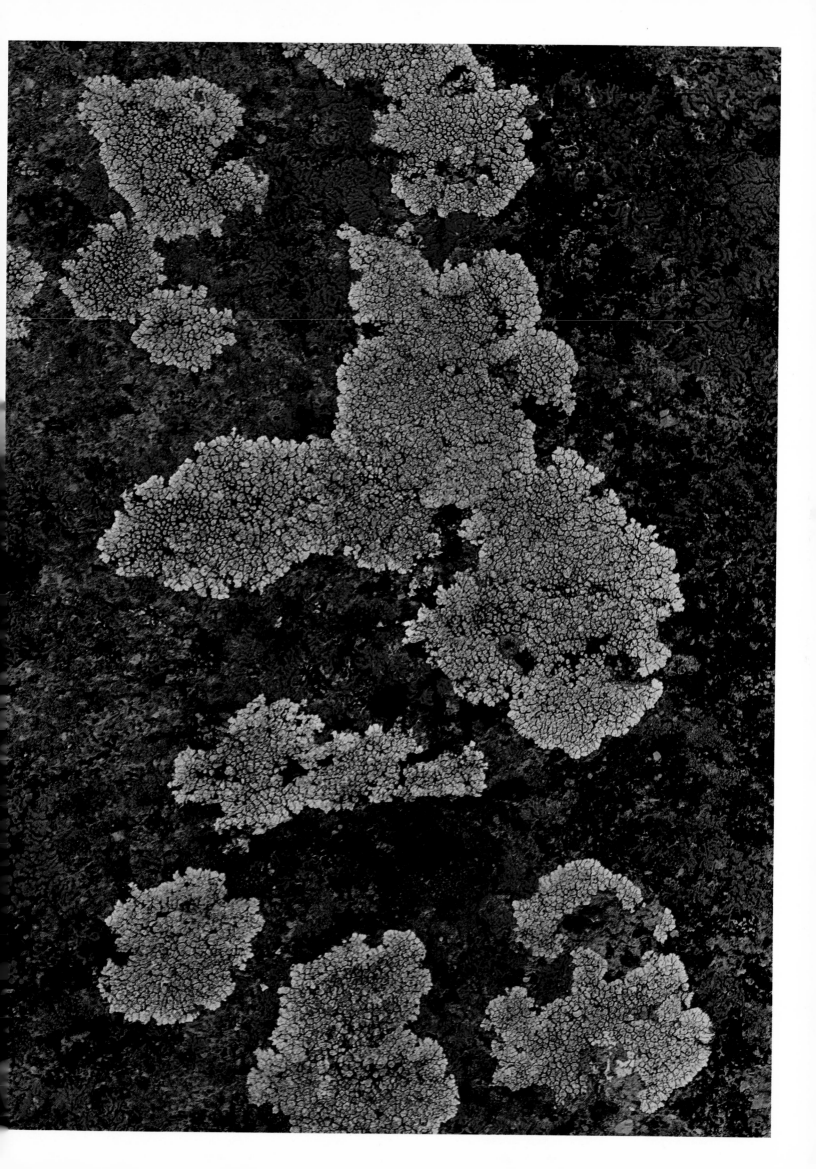

ecology of lichens: a knowledge of their relations to the world in which they live and the insight that this knowledge gives me in understanding the world that mankind lives in.

After we learn that rocks are more to man than building materials or paving stones; as we learn that soils and plants support us from a surplus that Nature normally allows them to manufacture; as we discover that animals cannot stand alone, we have just scratched the surface of natural knowledge. Then our hobby has become our education; it is in a position to lead us toward insight into the world.

When we find that rocks and soils and plants and animals do not live together in glorious isolation but rather that they affect one another, and that the effects produce other effects, we are studying the living world ecologically. We then study lichens and other plants not only for "themselves" but also for the parts they play in their own lives today and in the generations of their kind that follow them. We see more clearly their many services in the lives of "consumer" organisms, including man.

If ecology were just the cataloging of an unchanging world, where everything went through neat cycles, where everything had a role that it never outgrew, the ecologist might become competent—but bored. No such unhappy event will occur, however, for the drama, the roles, the players themselves, are always changing in one or another of Nature's great rhythms.

The "balance of nature" is not like the running down of a wound-up clock. The natural community in a very real sense not only rewinds itself, but rebuilds and replaces the clock as it does so. Such changes are not altogether haphazard, either: they are often orderly and more or less predictable. Thus, you can imagine the kinds of plants—not only the species of plants, but the general shape and life habits of them—that will flourish in a pond as it fills, first with water, then mud, then firm soil. You can see the pictures emerge and gradually change, as the water first has its free-floating plants, then its bottom-rooted submerged plants, then those that thrust their stiffened stems into the air. Finally, soil builds up, the water level is left behind, and marshes give rise to meadows and finally even a forest may evolve.

The change is ceaseless, it is gradual, it is orderly: it is the process known as *succession*. There is hardly another more important thing that can be learned about living things and the lives they lead.

Succession goes on in ponds that are filling; it occurs in stumps or old logs as they fall into disuse, disintegration, and decay on their way to becoming soil. Much the same pattern of events takes place in a dropping of manure in the meadow, a cup of water in a knothole, or in an abandoned field retired from the plow. In all these cases, the trend of changes seems toward some final, more or less settled state. In this "climax" condition, changes still occur, but normally they are the stable rhythms of maturity: the steady income and outgo, the quietness of maintenance; sleep and wakefulness; light and darkness; balanced manufacture and consumption. Such a natural community knows where it is going. Such a natural community is a perennial feudal society, with all the rich tapestry of contrast and vividness that feudalism implies. Deaths of individuals occur in it, of course, but there can be little doubt as to successors. Only catastrophe in the form of climatic change or overwhelming human interference or earthquake can destroy its relative equilibrium.

Lichens and certain other small plants enter very early into the ecological succession of an area. They are the great "pioneers" of the green world. They can survive in the driest places; they can eke out livings on the most exposed boulders of sandstone or granite. Such rocks would not ordinarily contain many minerals to feed plants or animals even if they were entirely ground into powder. Yet lichens are able to utilize the scanty atoms of nutrients that they can absorb from these unyielding rocks. They can certainly be called pioneers! They conquer; they live in

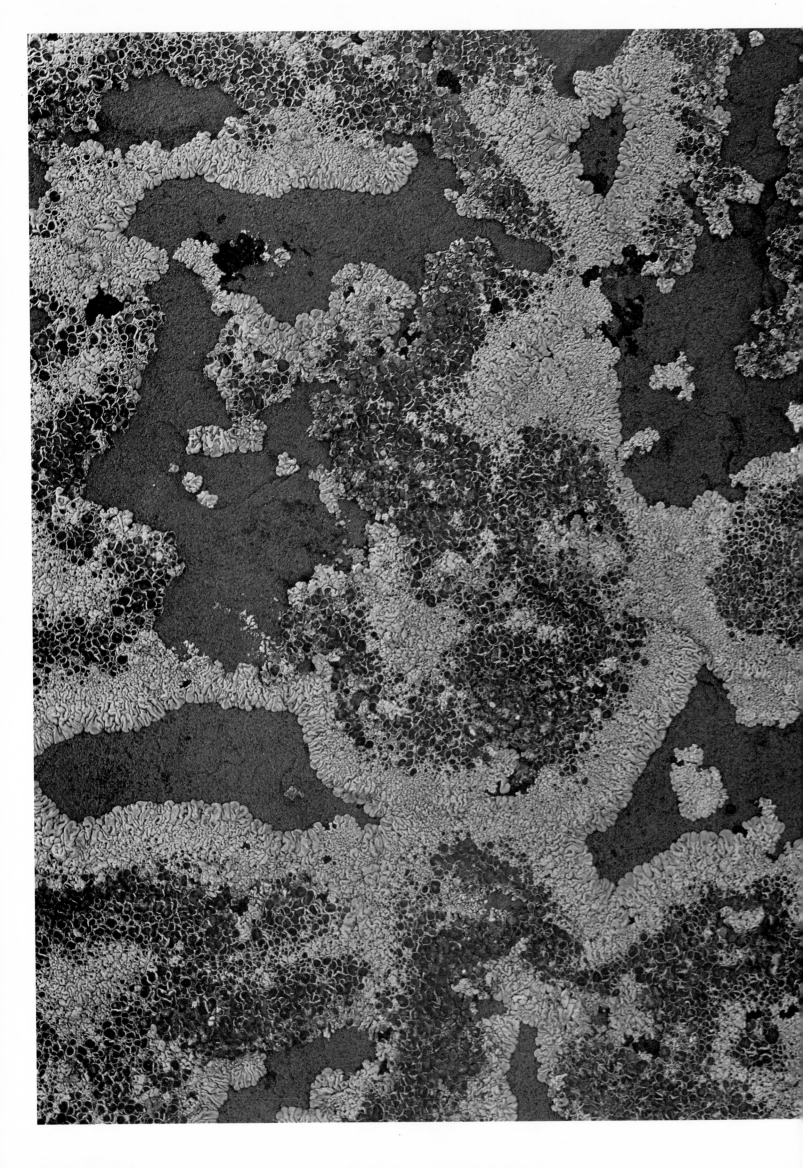

a place for a time; they add their bodies to bits of rocks that they themselves have pried loose. They prepare the way, and then disperse to new frontiers to colonize, to subdue.

Lichens commonly grow on very acid or "sour" soils. This means that nutrient minerals such as calcium, potassium, and nitrogen may be extremely scarce. Lichens are thus able to live on the bark of trees, bare stone, burned-over land, in cold, in waterlogged soils, in the Far North.

Red granites of great age occur in a few areas of the oldest part of the Missouri Ozarks. Near lovely Big Spring in Carter County, the top of an ancient mountain thrusts through the normal country rock of that area as a small outcropping of granite. One can go there on a summer's day when heat waves dance and cicadas drone, and walk among great rounded granite boulders that look like elephants humping amid the shortleaf pines. It is a dry and poor soil. It is the home of pincushions of moss and crusty rosettes of lichens that prefer—or at least succeed handsomely in—just such dry and exposed places. The beautifully rounded domes of granite wear a greenish frost of these rosettes; their sides are whiskered roughly with beard lichens. And at the foot of the huge rocks lies the abundant rubble of stone that has failed to withstand the assault of wind and rain and freezing and the prying attentions of plants. The bits of fragmented granite are on their way to becoming soil: the same thin soil now cherished by shortleaf pines, leadplant, New Jersey tea, and low, tough blueberries. In fact, in some places, only such hardy frontiersmen as pincushion moss and reindeer moss (*Cladonia*) can lay claim to survival on it.

Yet the beginning of soil formation here does not take forever, as some persons believe. Weather attacks the rock surfaces as fast as they are exposed. Indeed, weather-dominated influences begin to attack the rocks even before they are uncovered by normal erosion processes. And the lichens forge ahead. One sees new surfaces that were cut no more than twenty-five years ago (these rocks were once quarried), and they are wreathed with the lovely jade-green laces of lichen rosettes. (By establishing a growth rate of lichens by photographing them at intervals, clever ecologists have recently begun to construct "lichen calendars" for various parts of the Earth; by means of these measuring sticks, they plan to decipher the age of exposed areas.)

Due to soil acidity, and to the absence of minerals that commonly goes with it, lichens are very plentiful in sandstone country, too. Something that ought not to surprise even the amateur ecologist is that many flowering plants are also the same on the two kinds of soil. Thus in areas of southern Missouri where both granite and sandstone occur only as isolated outcrops, red maple may be practically restricted to soils derived from these rocks.

Lichens, as pioneers, do not make all their contributions to the natural community in a passive way. Acids secreted by their bodies help in dissolving rocks into smaller and smaller particles. Rosettes of lichens commonly deteriorate at the center as old age comes. This open, relatively protected area may be colonized by mosses or even drought-resistant plants of other kinds, if a little organic matter can accumulate to feed them and to hold a bit of moisture. Thus the frontiers of life expand.

A group of lichens that may even precede the rosettes lies in flat crusts on the surface of rocks, like delicate traceries on the face of the stone itself. They are living plants, gnawing their way

57

into the rock's surface and adding splinters of stone to the sum of broken-up things, as well as their own bodies when they die and decay.

Various species of beard lichen (*Usnea*), particularly lovely in their grotesque way, strike out to raise themselves off the stone's lowly surface. Spruce trees on the coast of Maine may be turned into pale ghosts of themselves by clumps of one sort of this slowly enveloping lichen. The sparsely leaved tops of Missouri oak and hickory trees are often festooned with another species. Yet another sort grows luxuriantly as long, silvery tresses on red cedar trees that hang grimly onto the great limestone bluffs along the streams of southern Missouri, looking rather like the South's Spanish moss, as many people mistakenly take it to be. But for beard lichens in treetops, you must either put on your climbers or wait for a tree to fall.

Some lichens grow on the ground's surface, of course. Reindeer moss (*Cladonia*) grows abundantly in this fashion on many barren soils in Missouri, as well as in the Arctic where it is so common as to have a popular name. As forage for man's livestock, of course, Missouri's reindeer moss is poor fare. But do not forget that it helps to keep the soil in place. This is important. The soil might one day become deep enough and filled enough with organic matter to be useful to us. And, obviously, anything that plays its part, however small, in the natural scheme of things *is* useful. If lichens do no more than hold thin soil on a hillslope and prevent its washing into a streambed, they have served a useful purpose, even from the viewpoint of the most "practical" person.

But lichens are beautiful in themselves. They need no justifying price tag. Tussocks of pincushion moss interspersed among tufts of the many-branched, silvery green *Cladonia* or the scattered clubs of fire-red British soldiers: these are as lovely as any plants can be. One of the greatest uses that a thing can have in this overcrowded, overadvertised, overhastening world is usefulness through beauty.

My hobby-interest in lichens is as a model. In their diminutive forms and lives, I see reflected the ponderous workings of ever larger and more complicated cycles of the universe. From atom to molecule to lichen to the tree that frames the airy bubble of the sky—and the sky itself—the wonders of nature are composed of wonder and productive of yet other wonders. The rosette of the lichen is a rose window of nature whose symbols lie openly beautiful to the eye but with meanings that are hidden. These symbols are truly magical virtues, for they tell greater stories than their physiology alone signifies. They are like a short, lovely poem that easily outlasts the closely reasoned and many-worded books that inspired it. By this I do not mean to imply that as a hobbyist I have become a Doctor Faustus with unearthly powers over matter and man. But the man for whom the chant of politicians sounds hollow and old, the person who wants to avoid treasuring the merely huge, and the one who does not want too quickly to worship the artifices of man: to them, I recommend the lichen.

"Lichens are the minutiae of the plant world.
They are not invisible, yet they are not large enough
to have lost the elfin qualities that go with perfection
in small packages. And to diminutive size they add
the charms of unexpected color and exquisite symmetry."

Yet Volcán Alcedo is tortoiseland supreme. Its inhabitants of the subspecies *vanderburghi* number perhaps 4,000, the most of any of the surviving races, forty percent of *all* the giant tortoises of the Galápagos.

Their home measures thirteen miles across its base, with an immense caldera that is four miles wide and a thousand feet deep. Steaming fumaroles and yellow sulfur castles are a subtle reminder that Alcedo is very much an active volcano, and last erupted only twenty years ago.

To the caldera of Alcedo, with camera, came a remarkable young woman whose Belgian parents moved to the Galápagos Islands when she was two years old. Educated by her parents, fluent in four languages, an authority on Galápagos natural history from years of leisurely exploration, and an extraordinary photographer, Tui De Roy spent eleven days in the realm of the tortoise, recording their shuffling life. These pages only sample the result.

The tortoises of Volcán Alcedo carry the dome-shaped carapace and have the short neck typical of those races that live where vegetation is relatively lush and easily reached. On arid islands, tortoises have evolved saddleback shells and long necks so they can stretch to reach the sparse foliage of shrubs and the fruits and pads of treelike prickly pears.

Long ago a privateer described the tortoises of the Galápagos in the most uncomplimentary words: "The creatures are the ugliest in Nature, the shell not unlike the top of an old hackney-coach, as black as jet; and so is the outside skin, but shriveled and very rough."

Perhaps, in utterly literal terms. But come with Tui De Roy to the caldera of Alcedo. Climb the pumice slope under the faint glow of a quarter-moon. Stand on the rim and listen to the singing of thousands of Darwin's finches, their voices carried on the wind from the floor of the collapsed crater. Camp with Alcedo's ancient reptiles and rise at dawn in the bottom of a volcano drowned in fog. Wait with the tortoises until, as the sun rises and the air warms, the cloud mass starts to pour out over the rim.

Now, as it has every revolution of the Earth for hundreds of centuries, the day of the galápagos begins, and you are an observer. And as you come to a better understanding of the ways of evolution, you are certain to question the parameters of "beauty."

In a Dantesque scene, Galápagos hawks wait atop sulfur deposits in the Alcedo caldera.

69

Kipahulu–Cinders to Sea

PHOTOGRAPHS BY ROBERT WENKAM

TEXT BY PETER MATTHIESSEN

A HIGH clear morning, after days of unseasonal rain. On the west rim of Haleakala, at over ten thousand feet, we are well above the clouds, but clouds still cling to the windward slopes beyond the eastern wall of the volcano. The east wall rises in black jagged silhouette against the rolling whites, and at the base of the wall, ten miles away at Paliku, a glint of sun catches the roof of the ranger cabin where we will spend tonight.

Single file, we start a slow descent into the crater. In an air as clear and still as the air in the cool bell of a mountain flower, the pack mule, led by Terry, raises thin spirals of lunar dust, and its hooves ring crisp on the old cinders. The bowl of the crater is heaved up by furnace-red eruptions—mounds and cinder cones—and streaked by strange burns of white and orange, and gray lizard blues.

In the first fifteen hundred feet of the descent, there is no plant life, only the cold fire bed of twisted stones. Then small spurts of the heathlike *Styphelia* appear, and stray dandelions, like puffs of gold on the black cinders, and the first silversword, ahinahina: on the steep sides of the cinder cones, the silversword is often the only life, shining like snow patches in a clear March wind. The silversword is a rare composite with a base of sharp silver leaves; it may grow for twenty years before bringing forth its mighty inflorescence, a column of red-purple flowers that can be taller than a man. Having done as much, it dies. A number of silverswords were in full flower, and here and there lay silver skeletons of others, dead in the previous year.

The west and north faces of the crater wall are bare as slag, but the low south wall is faced with a green wash of low brush and bracken fern; the bracken increases along our path as we drop to seven thousand feet. Here the crater floor levels off into a gradual decline toward the southeast, and fields of bracken, like a crop, rise from the black ash that flows in ancient rivers through the round red hills. The dandelions gather, and the first patches of endemic bunch grass, *Deschampsia*, but the ground cover is broken unaccountably by broad patches of desert emptiness, as if the crater here had not yet cooled. We reel across the emptiness like survivors, the dust of our passage rising like some lost signal into the ringing blue.

70

Dawn, Kipahulu Valley . . .
"This narrow sliver of the rim, only a few feet in width,
was reached at just that moment when the sun rolled through
the mist; as I cower here against the rocks, the crater
is one thousand feet straight down on our right hand,
and the Kipahulu a comparable distance straight down
to the left. Mist washes past us into Haleakala,
on a cold sea wind. The mist is touched by rainbows,
and from below come stray calls of the nene. A flight
of apapane, set free from the drenched foliage by the sun,
dance through the pass and drop away into the canopy
of cloud forest at the head of the Kipahulu."

"On the west rim of Haleakala, at over ten thousand feet,
we are well above the clouds, but clouds still cling
to the windward slopes beyond the eastern wall
of the volcano. Single file, we start a slow descent
into the crater. In an air as clear and still
as the air in the cool bell of a mountain flower,
the pack mule raises thin spirals of lunar dust,
and its hooves ring crisp on the old cinders. The bowl
of the crater is heaved up by furnace-red eruptions
—mounds and cinder cones—and streaked by strange
burns of white and orange, and gray lizard blues."

The blue is empty; there is no sign of a bird. On the wind from the high crags that march along the curling clouds comes the bleat of feral goats; goats have run wild in the Sandwich Islands since the time of Captain Cook. In company with the wild pigs that came with the first Hawaiians, the goats are held accountable for much of the destruction of the steep native forest, and are shot on sight.

Eastward, down the slow incline of the crater, the *Styphelia* brush becomes mixed with a true heath, *Vaccinium*, and we pass mamani, a pea tree valued for its dark hard wood, and the only sandalwood tree that has taken seed on the crater floor. Composites and a blue-flowered mint appear, and scattered small mushrooms, and some orange moss in the shadows of black rock.

The southeast corner of the crater rim has fallen, leaving an enormous breach, the Kaupo Gap. Here a torrent of dead lava rolls away into the sky, and clouds rise from below like somber mists from a great waterfall in the inferno. Above the clouds, far away on the far side of the channel, loom shadows of Mauna Kea and Mauna Loa, the volcanoes on the island of Hawaii.

The Paliku Cabin lies in the shadow of the eastern wall, at an altitude of 6,400 feet. The wall is very steep, and at its rim, swirling clouds constantly threaten. At Paliku, the annual rainfall is two hundred inches, over three times as much as it is in the open crater only three miles to the westward, and the cabin rests in a grove of native rue, with scattered mamani, olapa, and ohia-lehua trees; the ohia-lehua, a myrtle, is the dominant tree of the mountain forests of Hawaii. All these trees are rather stunted, and crusted over with lichens, mosses, shield ferns, and other epiphytes of wet terrains. South of the cabin, a rich pasture of *Deschampsia* flows off southward into the sky at the Kaupo Gap. Nene geese, once almost extinct, are raised here in a pen and encouraged to return to feeders in the pasture. Stripping my hot boots I walk down barefoot through the cool mountain meadows to have a look at them, and listen to their gentle calls to the wild birds beyond the pen. At midafternoon, the sun still holds the clouds at bay, and a three-quarter moon rises mysteriously out of a cleft in the mountain rim that we will climb tomorrow. Somewhere between my eye and its white ellipse is a hurtling silver thing in which the first men on the moon are coming home.

Toward twilight, as the crater cools, the clouds slip forward past Kuiki Peak, sending dank feelers down over the walls; the air cools rapidly, and we light a fire. In the night, the clouds settle into Haleakala, the House of the Sun, shrouding Paliku in a dense weeping mist.

After daylight, drinking coffee, we wait for the drizzle to ease. The mule will stay in the corral at Paliku Cabin; Larry Guth, the park ranger, will take it back tomorrow. From here we pack our equipment on our backs—a tent and fittings, sleeping bags, rifle, a machete, cameras and binoculars, a big tin container of dried food, a bottle of bourbon, and a change of clothes. All this is stuffed into three big burlap sacks and the sacks bound onto packboards with canvas shoulder straps.

At 7:30, the clouds still sink into the crater. In the mist, we cross the meadow and push through a thicket of Hawaiian raspberry to the foot of the escarpment; here a goat track ascends the cleft at what must be a sixty-degree angle. The track is mostly hidden in waist-high amaumau ferns, bright red and green, and we are drenched from the beginning. The laborious climb up greased mud and roots and around loose rock abutments, thrown off balance by the bulky packs, takes nearly an hour, from the crater floor to a low saddle on the rim under Kuiki Peak. The outside face of the narrow rim is the head of the Kipahulu Valley, a great forested ravine several miles across that descends all the way to the sea; the Kipahulu, which is our destination, is the last great stronghold of the Drepanididae, or Hawaiian honeycreepers, a family of birds even more remarkable than Darwin's finches, in the Galápagos, as an example of adaptive radiation.

This narrow sliver of the rim, only a few feet in width, was reached at just that moment when the sun rolled through the mist; as I cower here against the rocks, the crater is one thousand feet straight down on our right hand and the Kipahulu a comparable distance straight down to the left. For a moment, both walls of the valley are visible, as well as the central escarpment that separates the valley floor into two levels; each level has its own wild river, the Palikea under the northeast wall and the Koukouai under the south.

Mist washes past us into Haleakala, on a cold sea wind; this is the trade wind out of the northeast. The mist is touched by rainbows, and from below come stray calls of the nene. Around us, the black volcanic rocks sparkle with long tufts of silver lichens, and crimson vaccinium berries glisten in wet crannies. Apparently, the pass is used by birds of the high forest on both sides: a flight of apapane, set free from the drenched foliage by the sun, dance through the pass and drop away into the canopy of cloud forest at the head of the Kipahulu. Swirling mist, black crags, glistening trees, wind, sun, and rainbows: the setting for the first drepanids I have ever seen could not be improved upon. Moments later a green amakihi spun upward to a point of sun on a lichened branch tip, then folded its wings and fell into the forest like a stone. A note of red in the shining morning leaves of an ohia tree that overhung the precipice was a male apapane; it followed the rest into the green below. The pleasure I feel in seeing these drepanids has little to do with the knowledge that over half of Hawaii's native land birds are extinct, or nearly so (the island of Maui appears to have lost less species than the other islands, and the Kipahulu Valley is the wildest region of Maui); it comes from the wild light and the dancing birds and the two silent valleys.

Clouds come and vanish. We climb into our packs again and clamber up the hogback rim toward Kuiki Peak, where the south wall of the Kipahulu begins its long descent into the sea. I pay close attention to footholds and handholds, and a very good look at the subvascular flora is obtained. Two lichens of this wild high place are unlike any I have seen. One crusts the bark of the dwarf ohias that serve as handholds on the flank of Kuiki; it looks precisely like a seabrown algae speckled with gold-green dots, like fern spores. The other is leafy, silver-white with a vivid black border given it by its black underside. *Cladonia* lichen is everywhere, and sphagnum.

Erosion by wind and rain and goat has reduced the crest of Kuiki to a barren volcanic rubble. On the far slope of the peak, stray goat shadows in the mists, and Terry sheds his pack and drops down behind the rocks, clutching his rifle.

Now it is midmorning, and we wait for the weather of the day before—early mist and rain followed by clearing along the rim—but our hopes are vanishing. The crest of Kuiki is as isolated as a sea rock in the blowing cloud and rain. We huddle in soaked clothes behind an outcropping, waiting for Terry, who is waiting in turn for the clouds to pass so that he may kill his goat. The Linds, who are mountain goats themselves, do not believe in rain gear, and I am badly equipped because I am traveling light; I wear a borrowed jacket that no longer resists water. It is hard to believe that one could be cold in Maui in July, but I am shivering like a stunned fish.

In two hours, Terry returns, as silently as he went. He is a very tall, strong boy who is cheerful

74

"On the steep sides of the cinder cones, the silversword, ahinahina, is often the only life, shining like snow patches in a clear March wind. The silversword is a rare composite with a base of sharp silver leaves; it may grow for twenty years before bringing forth its mighty inflorescence, a column of red-purple flowers that can be taller than a man. Having done as much, it dies."

Through a screen of ohia lehua, the Kipahulu Valley
—looking across to the heights of Puu Kuiki.

but says little; in this, he is like his father, who is round-backed and steady as a barrel. Jack squats stolidly in the wet murk, nursing his canned heat; we gulp bouillon and crackers but find no warmth in them, and set off along the valley rim, in slow descent.

The descent from Kuiki to the sea is described after the fact; the rain and mud made it too difficult to take notes as I went along. That night I scribbled blind notes in the dark—to save weight, we carried no light of any kind—and the following night I did the same, and made what I could of these scrawlings in the days that followed.

We had agreed to go down the south ridge of the Kipahulu rather than descend the valley floor, which was buried in clouds; also, an expedition guided by Lind had come up the valley two years before, whereas nobody had ever traveled the south ridge, which presented us with an opportunity for exploration. We picked our way along its rim, in a thick mist. From the head of the valley almost to the foot, the rim remains about one thousand feet above the valley floor, and the fall is nearly vertical all the way. Now and then the clouds would thin to give a dim glimpse of the mountainside, but we depended for our bearings on the edge of the precipice. Not that there was much chance of straying from the ridge unless one fell from it, since its far side was only a few hundred yards to the south, falling away into the Kaupo Gap.

Soon rocks gave way to alpine tundra, and we crossed a crystal mountain pool inset in a young stream, softened by mist. The tundra was inhabited by the dark twisted trunks of solitary ohia trees: the flowing clouds, thinning and thickening, gave the place an eerie transience, as if we were passing from one world into another. Jack Lind, who had never seen this corner of the mountain, was enchanted; he kept stopping to peer about him, shaking his head. "I'll come up here and stay for weeks," he said, "if I can get Terry to come with me." Originally we had planned to make our high camp in the grassland, but the poor weather drove us off the mountaintop. We had come down perhaps a thousand feet, and imagined we might make that point on the south rim, at 3,800 feet, to which the Linds had slashed an upward trail in the past year.

At first, the going was merely uneven; we could not always go where only goats had gone before. Still, the goat trails down along the rim, honed to the most efficient routes by time, served us very well, and by midafternoon, another thousand feet of the descent—or so we estimated—had been accomplished. Furthermore, we had seen the first crested honeycreeper, which came to feed in the flowers of giant saxifrage. But it was just here that our doubts and troubles began. Due to the clouds, we had had only glimpses into the abyss, and once, when the mists lifted, Lind glared in surprise at what he could see of the steep ridge just opposite; it looked too close to be the central escarpment of the Kipahulu, and Jack, looking haunted, was momentarily convinced that in the mist we had wandered southward onto the ridge of another valley. But Terry and I, from our recollection of the maps, did our best to convince him and ourselves that this was not possible; we pitched ahead.

Just after the alpine grassland had been replaced by dense scrub forest, Jack had remarked that the valley itself was harder going than this ridge; now he changed his mind. The forest had become so tangled that even the goats had been turned back; for the rest of this day and all the next, the machete was in use full time. A break in the mist revealed that this south rim, at about six thousand feet, leveled out in a narrow peninsula of thick forest. We would be here tonight and a good part of tomorrow before a real descent could be resumed.

We chopped our way ahead. The forest was mostly ohia, interspersed with aralia, saxifrage, and rue; many of the ohia trees were prostrate, with dense ranks of vertical branches, like prison bars, and others had grown prop roots, to accommodate themselves to the wet terrain. The soft ground, hidden by rank ferns, was root-ridden and rotten in the gloom of a canopy that, in this weather, cut off all the light; it was a troll forest, squat and dank, all drippings and creepers and hanging shapes, picturesque in a weird way and quite impenetrable. Furthermore, the ridge had narrowed, falling away sharply on both sides, and presented a very small choice of route. Toward twilight Jack said, "We better find a place to camp," but we never came upon a clearing. Finally we settled for a spot where the fallen trees had not yet been replaced by live ones, and hacked out a less than level place that was just big enough for the small two-man tent. Fronds from the tree ferns made an acceptable substitute for air mattresses. While Jack and Terry pitched the tent, I set up a wind shelter for canned heat—a wood fire in this soggy place was not a possibility —and squeezed a potful of fresh rain from the mosses that carpeted the trees, then filtered it by pouring it through a neckerchief. Dissatisfied, Jack ran it through a yellowed T-shirt that he peeled from his own wet back to serve this purpose, but the water remained the color of bitter chocolate, so we put it on to boil without further ado. Some of the cruder elements surfaced in the pot in the form of a scum that could be scraped away, and such matter as remained gave body to the dehydrated soup, and did nothing to harm the flavor. Jack cooked the soup to a kind of lava in which thick bubbles bloomed and burst, and we ate it standing in the raining darkness. Then one by one, our dinner done, we hunched under the fly of the little tent and dragged off pants and boots in one amorphous mass and left them out there in the rain; we squirmed into dry clothes and got into our sleeping bags, warm for the first time since midmorning, and tried as best we could to make room for the others. Jack is broad and I am tall, and Terry has more big bones than either of us, and all three have strong shoulders; poor Terry, squashed in between, evolved a trick of locking his shoulders the way a bird on a limb locks its feet when it falls asleep, so that no amount of quick shifting or devious elbow work could shift him an inch. On the contrary, even in sleep he spread so rapidly into any vacuum that, turning back after lying on my side, I would find myself cockeyed, hung up on that implacable shoulder, my nose pressed into the wet canvas of the tent in which, Jack claimed three people could make do in a pinch.

All night it rained, and it was raining in the morning. Since dry clothes would be soaked in minutes, there was no choice but to reach out into the cold mountain mud for the rags of yesterday and wring them out as best we could—they were slippery with mud—and drag them on over bare quaking skin. By the time the wet pants were on, the rain was already pouring down my back, and I was shaking: I remember thinking, I'll never get warm again today, and I was right. There was no slow torture about it, as there had been the day before: we were racked with cold right from the start. Jack Lind was wearing the solitary rain jacket that we had between us, and perhaps because no one else had one, he wore it wide open, so as to derive the least possible benefit.

It seemed to take hours to drag the pathetic little camp back into the muddy sacks and lash the sacks onto the packboards. By eight we were under way, but our progress was too slow to speed the blood. Only the lead man, swinging the machete, had a chance to warm himself, and his advantage was offset by the cascades of cold water that the slightest touch brought down

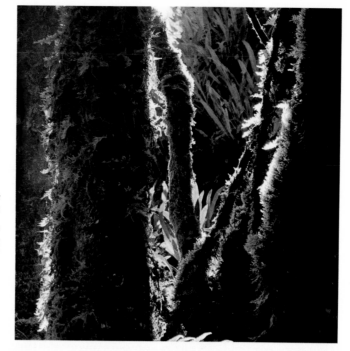

"*We chopped our way ahead. The soft ground,
hidden by rank ferns, was root-ridden
and rotten in the gloom of a canopy that,
in this weather, cut off all the light;
it was a troll forest, squat and dank, all
drippings and creepers and hanging shapes,
picturesque in a weird way and quite impenetrable.*"

*"In the early afternoon, we came upon an old machete cut, though the trail
had been grown over. We were off the level ridge now, and descending steeply.
At just after four in the afternoon, we came out on a grassy point. Here the sun
shown, and the mist lifted briefly to reveal the blue Pacific thirty-eight hundred feet below.
We sank down in the wet grass and had a drink. I was footsore and stiff-legged
as a zombi. In eight hours of violent hacking and falling and crawling
and climbing down this godforsaken ridge, we had descended at most
two thousand feet, and covered a distance of less than two miles."*

from every branch. Each time we dragged ourselves over a fallen tree, the spongy moss soaked us all over again. More than once we were belly down in mud, bumping along under the hairy trunks, the limbs catching at our load, but already we were too begrimed to mind the mud; the enemy was cold. By noon, no one suggested that we stop to eat: in the blowing rain and mist that swept up out of the valley, it was not bearable to stand still long enough to fumble nourishment out of the shapeless packs.

Now and then during the morning came glimpses of small fleeing birds, mostly amakihi and Maui creepers, and one iiwi, and a crested honeycreeper (the day before we had seen three honeycreepers, which are orange-streaked black birds with a curious "crest" at the base of the upper bill; these akohekohe are now restricted to Maui, and even here they are considered very rare, though they are not uncommon in the Kipahulu). But in the dark dull light, the colors of the birds were lost; but for Jack Lind, I would not have known what I was looking at.

Here in the cloud forest, the epiphytes had taken over, not only the mosses and the ferns, but iris and mistletoe and plants such as *Styphelia* and the tree ferns and the astelia lily, which had been entirely terrestrial at higher altitudes: apparently the ground in the lower forest is too wet for them. The aralia, rue, and saxifrage families all occur here as small trees, and also several varieties of giant lobelia. On the end of the twelve-foot stem of one lobelia was a lovely lavender inflorescence—the only nonchlorophyllous color that emerged from the green monochromes of mountain forest. The woody lobelias of Hawaii are one of the most striking features of a unique flora, and there are at least twelve species in the Kipahulu alone. This strange valley should be protected if only for the fact that nearly ninety percent of its higher plants are native, a situation which is unique in the islands. "Within three miles," wrote Dr. Charles Lamoureux, the botanist with the 1967 Nature Conservancy expedition, "one can find communities ranging from tropical rain forest to a subalpine zone with frequent frosts. . . . Since most of the Hawaiian species of plants are endemic, these communities are like no others, and Kipahulu in this sense offers an opportunity [for research] not available elsewhere on this planet."

But I remember best how little I noticed; one needed one's full attention to deal with the constant obstacles, and anyway, there was simply too much rain. Falling repeatedly, or catching the feet on hidden roots, or jamming a leg into mossed-over holes between the rotting carcasses of fallen trees, or nicked by a deflection of one's own careless frustrated machete, I was very conscious of how helpless a man would be with any disabling injury. I said as much to Jack, who merely grunted, shaking his head: he didn't want to think about it. A stretcher could not be carried through this tangle, which was also inaccessible to a helicopter. In consequence, one

paid close attention to the footing, and dealt with the obstacles right before one's face; there was no margin for accident, much less error.

In the early afternoon, we came upon an old machete cut, though the trail had been grown over. We were off the level ridge now and descending steeply. Not long after this, Terry shot a great bristly brindle sow, long-snouted and high-backed; the old Polynesian pig lay wide-eyed in a blood pool in the broad mud wallow of her own making. Jack slashed some meat off her tough ham, and we went on again. Glancing backward in the rain, I caught the eye of this wild-eyed old pig-of-the-mountains.

The trail was beginning to emerge, and we made progress. At just after four in the afternoon, we came out on a grassy point of the south ridge where Lind had cached some stores. Here the sun shone, and the mist lifted briefly to reveal the blue Pacific, thirty-eight hundred feet below. We sank down in the wet grass and had a drink. I was foot-sore and stiff-legged as a zombi, and Jack wore an expression of stunned surprise, as if a friend had kicked him in the stomach. "Now I know why nobody ever came down that south ridge before," he said. "You have to be crazy." In eight hours of violent hacking and falling and crawling and climbing down this godforsaken ridge, we had descended at most two thousand feet, and covered a distance of less than two miles.

At thirty-eight hundred feet, in the new sun, the air was warm. Lying in the dank grass, we finished the bourbon, staring in silence at the sparkling wet blossoms of the ohia trees and the white crescent of surf on the black shore of eastern Maui. From the valley below rose the thunder of the Koukouai Stream, now a mountain torrent. Far below, where the valley nears the sea, lies a string of paradisical waterfalls and pools known as the "Seven Sacred Pools"; although the remnants of an ancient Hawaiian temple lie nearby, the pools have been sanctified only in recent years, as part of an attempt to lure tourists to the remote, rainy, and immensely beautiful coast of eastern Maui.

Terry went to fetch water and Jack and I put up the tent. While the light lasted, Jack, lying on one elbow like a Roman, cooked up a great batch of soup and rice and dehydrated beef stroganoff, while I fried strips of the wild pig that Terry was carving from the ham. This was our first real meal since the supper at Paliku Cabin, two nights before.

The sun was still shining when we lay down in the tent, and we were much too tired to cut fern fronds. Jack Lind was snoring before my head touched the ground.

By morning it was raining once again, and once again we climbed into cold clothes, but the air was much warmer at this altitude. Jack and I broke camp, and Terry went on ahead. There was a discernible trail from this point onward, but now we were assailed by tropic downpours. The trail was slick, and at times it slithered along ridges so narrow and precarious that I was tempted to sink onto my knees and crawl, but my knees were so sore from clambering over the dead trees of the day before that I kept my eyes glued to the ground instead. Descending steep rain-slick ridges with a bulky pack is tricky work, and even surefooted Jack Lind had his feet go out from under him repeatedly; since he always landed on the food tin strapped to his back, he made a gloomy booming sound each time he fell. As for myself, I was flat on my back so often that I had ample time to contemplate the changing scenery. The ohia forest had been replaced by lovely great koa acacia trees, some of them six feet in diameter; farther down, the koa became

82

mixed with tropi
trees, and ti plar
places that had r
went along I nil
fiddlenecks of th
Hawaiians consic
guavas.

As we descen
noise like the oce
at three thousanc
by new invaders
which is frequent
nineteenth centu
only land mamm
the lovely tree sn
combat still ano
lower mountain i
cat, and commor
step without wate
tant sunlight, anc
the silent mounta

At one thousar
left a Land-Rover
the end in sight,
did was strip off c
get the mud out
shining sun, only
out of sheer relief
never been done
again."

*Waimuki Falls plunges toward the Pacific
—and the Seven Sacred Pools. Here the ohia
forest has been replaced by koa acacia,
some as large as six feet in diameter.*

Rocks of Lobos

PHOTOGRAPHS BY DENNIS BROKAW

TEXT BY DAVID LEVESON

THE SEAM where continent meets ocean is a line of constant change, where with every roll of the waves, every pulse of the tides, the past manifestly gives way to the future. There is a sense of time and of growth and decay, life mingled with death. It is an unsheltered place, without pretense. The hint of forces beyond control, of days before and after the human span, spell out a message ultimately important, ultimately learned: now, forever, mortality, infinity.

At Point Lobos the message is conveyed on many levels. Here, just south of Carmel, California, an intricately fingered peninsula juts into the Pacific. Its length and breadth are only two miles, and its highest point just a hundred feet above the sea, but within this tiny, fragmented extension of the continent, a thousand changing marvels wait.

More often than not, the peninsula is washed by fogs and mists and abounds in hidden places. At the coast itself, the crash of waves, the screams of gulls, and the bark of sea lions vibrate through the air. Where eroding cliffs provide a rocky platform, tide pools are filled with flower-like sea anemones, and hosts of shelled creatures adjust their shapes and geography in accord with the tides.

In its geometry alone, the line of the shore becomes a source of endless discovery. Following fast, one upon the other, sweeping curves culminate in jagged, dangerous points; keyhole-shaped bays evolve into narrow, elongated slots; straight coasts veer without warning into complex recesses or dead-end against overhanging cliffs. A minute's walk and the constant bite of open, roaring ocean is left behind. Where seawalls have succumbed to storm, placid, pondlike inlets are filled with floating kelp and lined by evanescent stretches of sand or pebble. The join be-

86

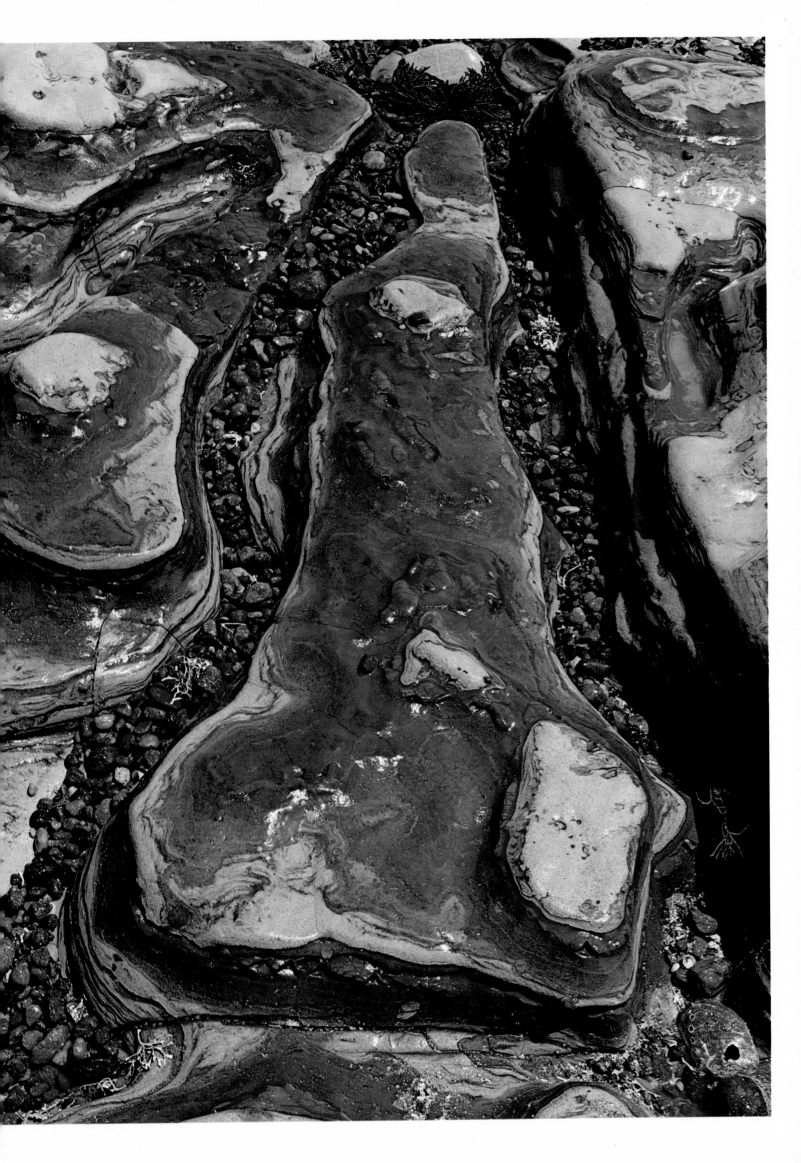

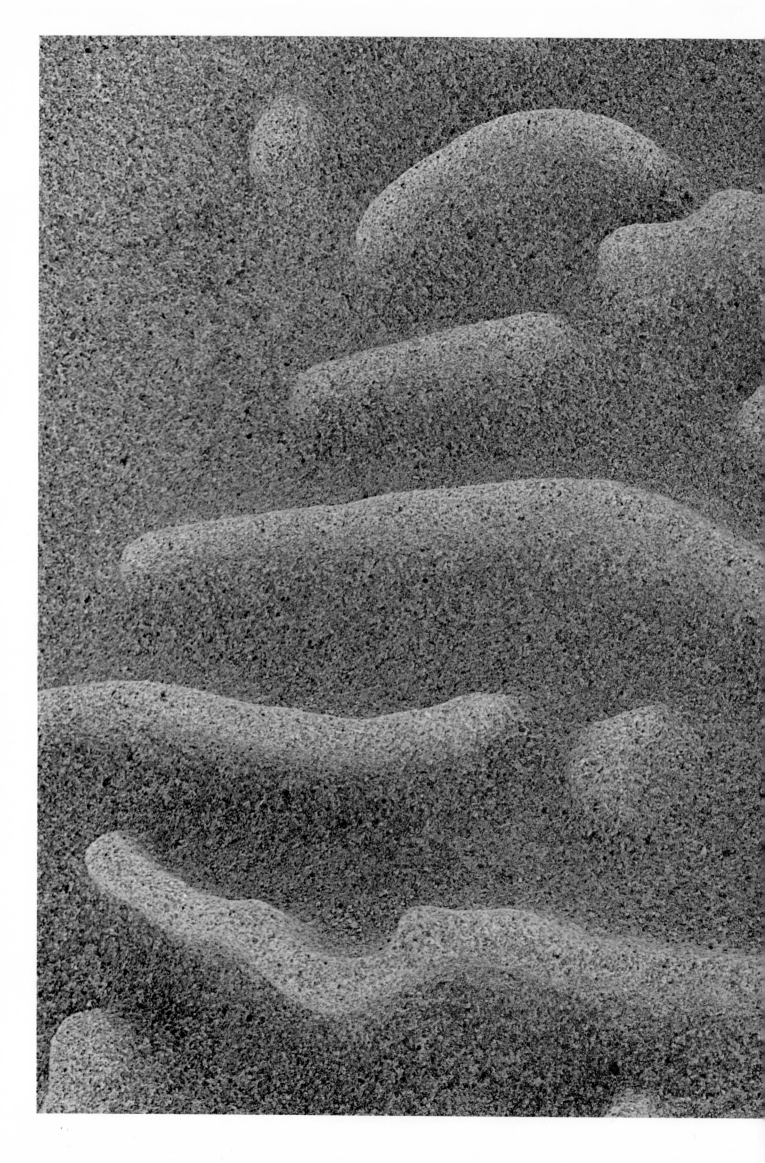

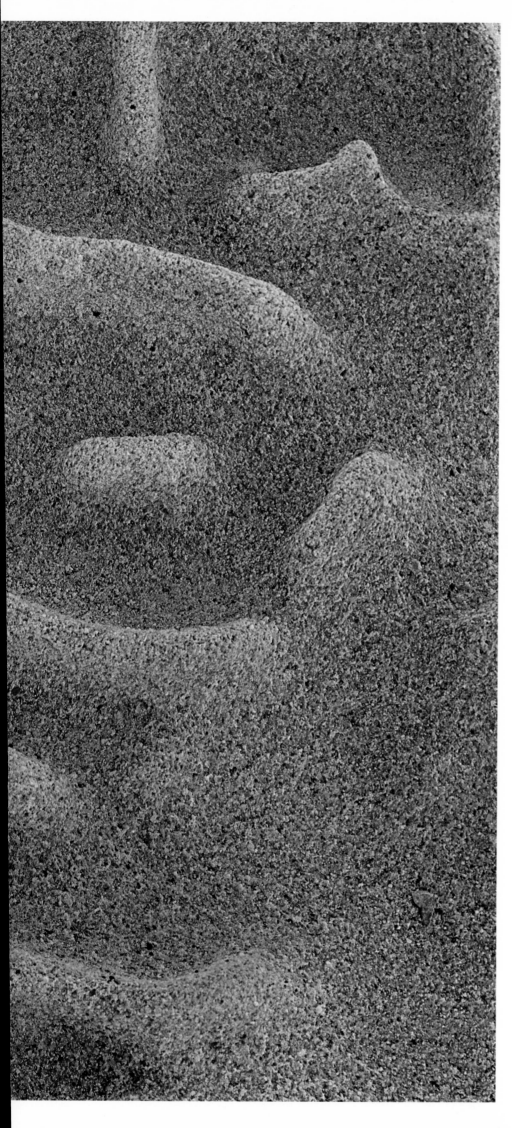

Half-inch-high, fingerlike
ridges, set in a matrix of
coarse, salt-and-pepper-
textured sandstone, mimic
the pointillist technique
of the French impressionist
painter, Georges Seurat.
The salt and pepper is a
mixture of quartz, feldspar,
and black mica grains bound
together by a natural cement.
Mica weathers and is removed
more easily than quartz
or feldspar. Thus wherever
mica is abundant, the rock
erodes preferentially, and
miniature valleys result.
The light, mica-poor areas
remain as miniature ridges.

89

A network of rock fractures in the sandstone has been enlarged by the erosive action
of seawater to form hollows that trap the tide and provide moisture for black turbans
—herbivorous snails that live between the tide's limits. Nearby, the asymmetrically
conical shells of rough limpets echo the curves of rounding promontories and elliptically
striped recesses. Yellowish rock layers are composed of coarse sand grains deposited
in the turbulent near-shore environment of an ancient sea. The gray, finer-grained layers
accumulated in quieter offshore areas where clay and mud could settle out of suspension.

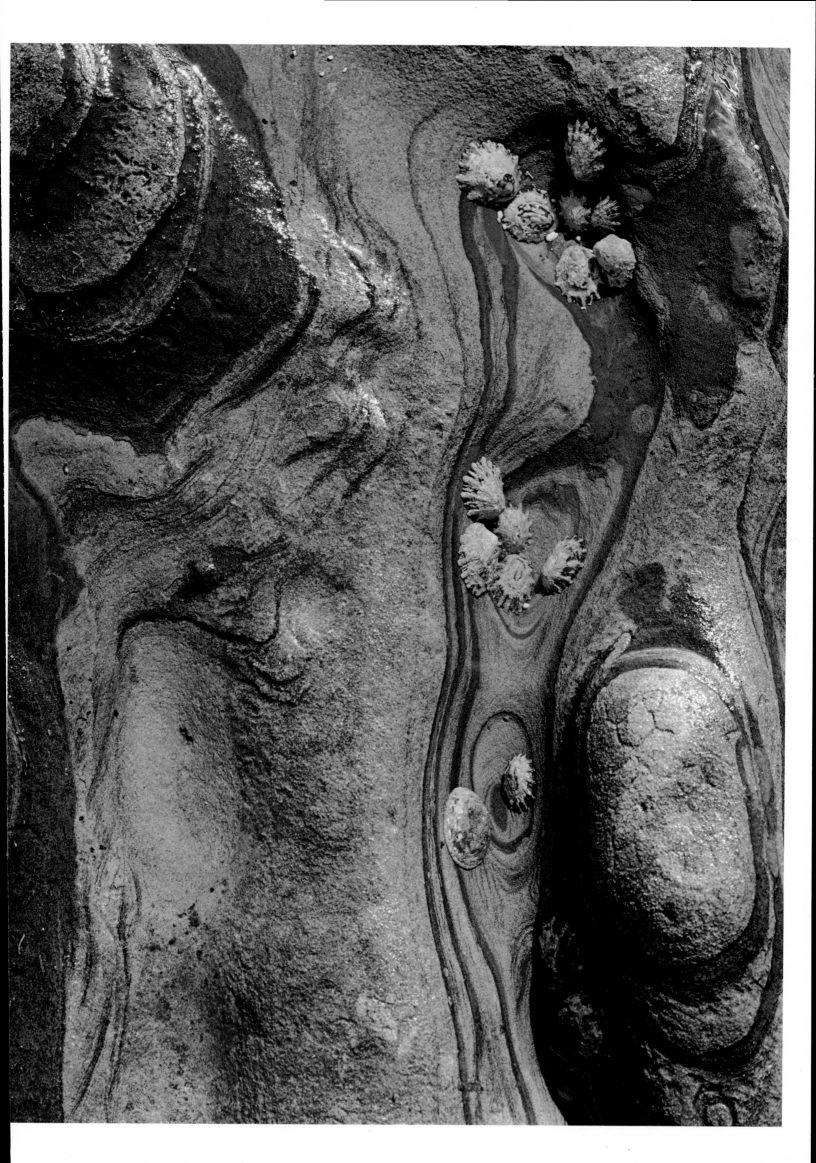

tween land and sea is organic and vital. Mind and body are steeped with a sense of the moment, of vision.

But this complex of process, ceaselessly absorbing, lies and works upon surface and material that in itself constitutes, multiplies, and extends miracle. Sixty million years ago, sand and pebbles poured off an ancestral California into its coastal waters, forming layers of sediment that became buried and gradually cemented to rock. With the passage of time, these rocky layers were caught in the jaws of a restless Earth, tilted, warped, and finally revealed once more at the continent's edge. Exposed to the corrosive pounding of salty water, the sting of wind-driven sand, the acid decay of organic life, all the varied substance and texture of the rock were tested in microscopic detail for strength and endurance, their separate characters accentuated and high-lighted in manifold ways as the weaker materials became decayed, detached, and washed away.

Today an intricate mosaic, alternately delicate, sensual, and grotesque, has evolved. Knobby, fist-high promontories give way to smooth, sculptured aprons. Armies of wavelike shapes, frozen in rocky stasis, are poised, seemingly forever, above expectant plains. Bulbous stone lumps sit Buddhalike in bowl-shaped cavities. Smaller still, rock fractures filled by resistant precipitates etch into inch-high ridges that stand above inch-deep valleys. Plateaus, only yards wide, are carved into a jigsaw puzzle of miniature canyons and divides, mesas and buttes.

The sense of scale shifts more strongly and the shells of tiny limpets and snails in the hollows become enormous, forbidding structures—the dwelling places of an alien, extraterrestrial race. The salt-and-pepper texture of sandstone assumes, in its structural variations, the magic of a painting by Seurat viewed at close range, the potential of other shapes, other meanings. Iron, manganese, and carbon in minute quantities stain the layers a subtle green, gray, orange, yellow. Concretions with visceral gleam and hue lie half exhumed, embedded on slabby, sandy plat-forms: earth organs, dissected, exposed, vulnerable, visibly palpitating, alternately shining and dull as tidal waters wash back and forth.

Point Lobos mimics, deceives, exhilarates, mocks. The geography of a continent is compressed into an acre, a square foot, the surface of a pebble, if one looks closely and carefully enough. It lends, momentarily, impressive size and stature to the human form. One grasps the whole rather than the part. One floats above rather than being immured within. It confers, for an instant, the illusion of omnipotence. Then the eye lifts and another message is clear. All that we are, all that we cherish, ends up here, at the join of land and sea, drifting, decaying, merging with its sources.

Under the constant stress of pounding waves and the corrosive wash of salt water, the sixty-million-year-old Carmelo formation at Point Lobos has been carved into a microuniverse of miniature canyons, mesas, and divides. The light-colored sandstone layers are relatively resistant to erosion and cap most of the ridges and summits; the gray-green shaly layers are more easily worn away and floor the valleys and slopes. Scattered reddish-orange patches mark local concentrations of iron.

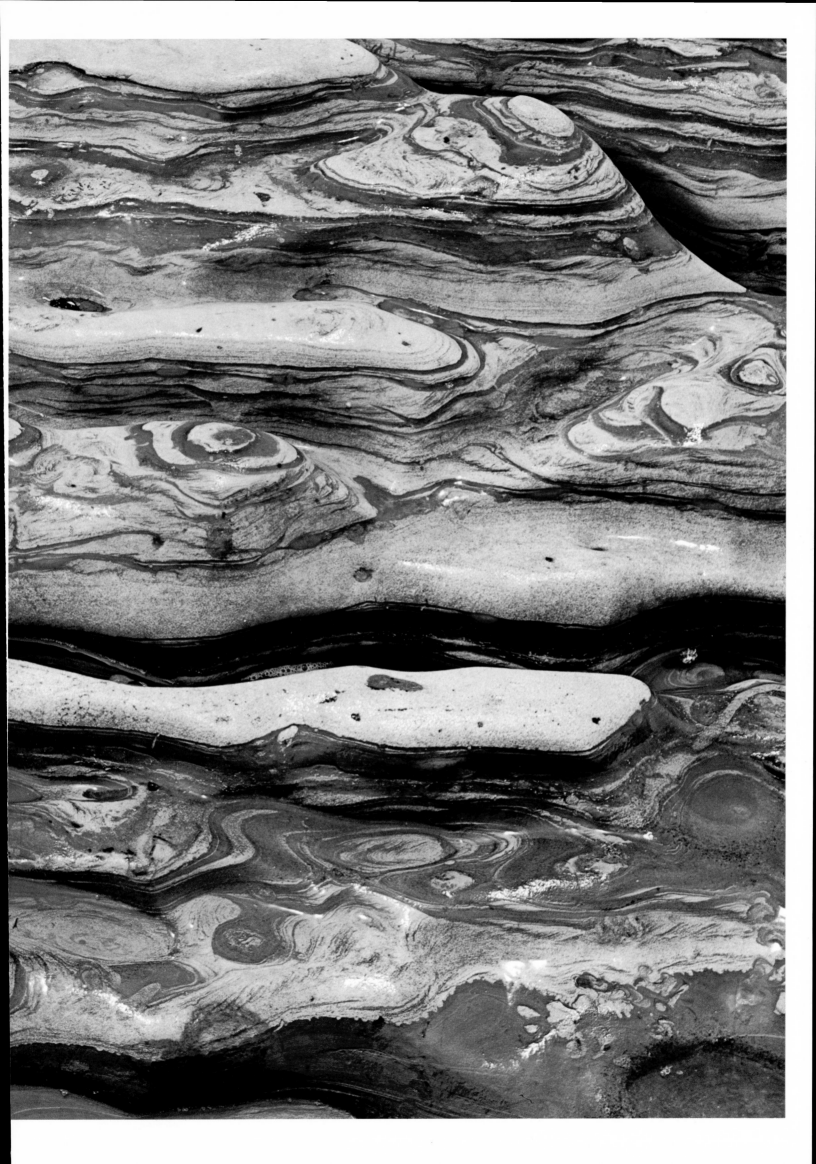

The jellyfish that drift on the current or pulse rhythmically about hardly resemble fish, and zoologists often call them "medusas." This is a reference to the mythical maiden with serpents in her hair, whose face turned a beholder into stone. Indeed, the tentacles of the jellyfish can paralyze their prey, and a few species pose a degree of danger to man. But terrible to gaze upon they are not. This is the lion's mane jellyfish, whose umbrella may stretch eight feet across, with tentacles that trail in the sea for 200 feet.

Life in a Cold Ocean

PHOTOGRAPHS BY DOUGLAS FAULKNER
TEXT BY LES LINE

THE ARDUOUS art of undersea photography has no finer practitioner than Douglas Faulkner. But like his contemporaries, Faulkner is lured firstly to the coral reefs, to those tropical waters that are populated by a gaudy and seemingly infinite variety of animal and plant life. His personal Eden is the Palau Islands—volcanic and coral-formed jewels in the Philippine Sea that support one of the richest of all marine faunas.

What is unfortunate is that diver-photographers, in their pursuit of the wonders of the reefs, have mostly ignored another oceanic world—that of colder northern seas—where life, although certainly not as lavish and diverse, is every bit as marvelous.

Thus *Audubon* sent Faulkner to four cold-water islands. To Amchitka Island, far out in the Aleutian chain; to Baranof Island, 1,800 miles away on the coast of southeast Alaska; to Vancouver Island, 700 miles to the south, largest island on the Pacific shore of North America; and to Santa Catalina Island, 1,200 miles farther south, within the limits of Los Angeles County.

The common denominator is a surface water temperature that ranges from a frigid 40 degrees in early summer at Amchitka, on the Bering Sea, to a more comfortable but still chilly 60 degrees off Santa Catalina. In contrast, reef-building corals cannot thrive where the surface water falls below 70 degrees, and the waters around a tropical paradise like Palau warm to 87 degrees.

Not surprisingly, Faulkner's first venture into a cold ocean environment was a physical and mental shock. The scene was Vancouver, the time November, the surface water and air temperature were both forty degrees, and it naturally gets colder as you swim deeper. "I came poorly prepared," he ruefully recalls. "My wet suit was too thin and had too many zippers that permitted easy entry—not only for me but for the icy water. I made eight dives a day, and I quickly came

Continued on page 113

94

Most damselfish, inhabit tropical seas.
They are colorful and small, less than 6 inches long,
and their numbers include several species that live with
impunity amidst the lethal tentacles of sea anemones.
But the brilliant-orange garibaldi is an exception
to two rules. It is 11 inches long, and it has adapted
to colder waters—the rocky coasts and kelp beds
of southern California. And it is pugnacious, defending
its carefully isolated nest in a clump of red algae
against all intruders upon its territory, including man.

They are the windflowers, the anemones, of the sea. There are a thousand species,
in every ocean from the poles to the equator, at every depth from tidal pools
to abyssal trenches. They are elegantly symmetrical, their hues ranging from pink
to orange to red, from white to brown to green, often with contrasting patterns
of dots and stripes. But they are carnivorous animals, and their "petals,"
like those of the lovely dahlia anemone of northern waters, are stinging tentacles
that cram food—from minute creatures to fish—into a centrally located mouth.
Yet anemones may have special "friends"—fish that dart among the tentacles
and lure other fish to their host, tiny shrimp that live in safety beneath the
deadly canopy, or even hermit crabs that carry anemones about on their shells.

A relative of the anemones is the quill-shaped sea pen, anchored in the soft bottom by an expandable, water-filled bulb at the base of its stalk. Sea pens, which reach three feet in height, trap plankton with rows of polyps along the barbs of the "feather."

Sponges were once thought to be plants. They are also plankton-feeders, straining sustena through millions of microscopic pores. A remarkable assortment of small creatures live and in sponges; this finger sponge hosts an algae-camouflaged crab and some tiny crustace

Surrounded by a colony of branched bryozoans, the empty shell of one crustacean
houses another. For a tiny shrimp peers from the opening where feathery feet
once pulsed in and out to kick plankton into the mouth of a giant barnacle.
The kinship of barnacles to the more familiar crustaceans is hardly apparent
except to scientists examining the larval stages. But such large barnacles,
which reach a foot in diameter, are as edible as crabs, prawns, and lobsters.

In search of the sea stars on which it preys, a hairy triton slowly crui
a submarine wall decorated with grapelike sea squirts and pink splash
of crustose coralline algae. Only 4 inches long, the hairy triton is fou
from California to Alaska. Largest of all tritons is the trumpet she
which Shinto priests in Japan use as a horn to summon their worshipe

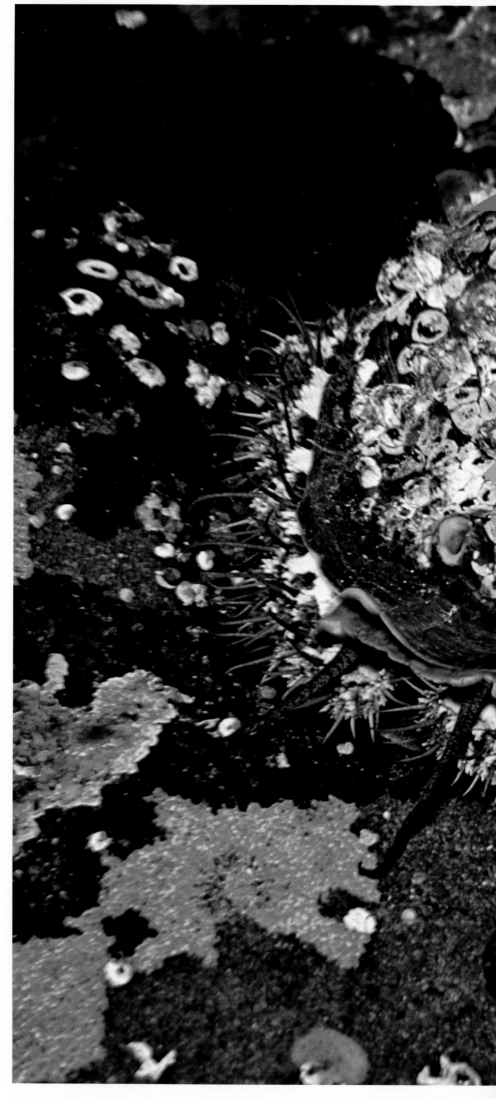

Bejeweled with encrusting life,
an abalone searches for food
with long sensory tentacles that
project from beneath its mantle.
Abalones are vegetarians, eating
kelp and sea lettuce; if a piece
of seaweed touches its tentacles,
the mollusk whirls and clamps
the plant against the rock with
its foot, using rows of teeth on
its "tongue" to shred the catch.
Holes on the abalone's shell
vent water that passes through
the animal's gills; the shell
itself may harbor a community
of marine life—algae, sponges,
barnacles, hydroids, bryozoans,
snails. Most abalones cling to
ledges as deep as 1,200 feet;
but others live in the surf,
grasping wave-pounded rocks and
feeding on microscopic plants.
An abalone will spend its life
in one area, and some grow to
a foot in size. Their enemies
include octopuses, starfish,
crabs, sea otters—and man,
who prizes both their flesh
and their iridescent shells.

*The clownlike face of the treefish belies its membership in a fearsome family
—the scorpion fish, whose venomous spines can cause painful and even fatal wounds.
More menacing in appearance is the sculpin, but its relationship to the colorful
treefish is quite apparent. Most dreaded of all is the stonefish, which resembles
a chunk of coral; to step on its spines means instant agony, and often death.
The spines of the scorpion fish are for protection; to capture their own food, they wait
for smaller fish to pass, lunge with blinding speed, and swallow their catch whole.*

107

There are some 250 kinds of sharks,
in warm waters and cold, and they
vary greatly in size, appearance,
and habits. There are docile giants
that feast on plankton, and monsters
that will prey on man. There are
beautifully marked little sharks,
and there is the awesome hammerhead.
The bottom-dwelling horned shark
is unique in several ways. Both of
its dorsal fins are armed with a
large spine—the horns. Its piglike
face results from lateral nostrils
and the most remarkable mouth in the
order of sharks. For the dental plates,
with their many tiny teeth, curve
upward from the upper jaws, downward
from the lower jaws. Thus it can dine
on crabs and hard-shelled invertebrates,
including mollusks dug from the sand.
Horned sharks may grow to 5 feet.

Sea stars, brittle stars, sea urchins, sea cucumbers, sea lilies, sand dollars—
these are the echinoderms, a term that comes from the Greek echinos, for hedgehog,
and derma, for skin. In other words "spiny skin," a description best suited
to sea urchins. The sea stars, or starfish, are perhaps the loveliest creatures
in North Pacific waters, and the variety of shapes, colors, and numbers of arms
is seemingly endless. They may have 5 arms, or 11, or 18, or even 24. They are
carnivores, feasting on mollusks, crustaceans, corals, and others of their own kind.

Terrestrial snails and slugs, abalones and limpets and periwinkles, clams and oysters, cowries and tritons, squid and octopuses—all belong to the second largest phylum in the animal kingdom. They are the mollusks, 80,000 species strong, found from mountain peak to oceanic valley. Their collective name comes from the Latin mollis, for soft; it is an apt description, whether the creatures own a shell or go about naked. Among the latter are the wide variety of nudibranchs that swim gracefully around, or creep along kelp fronds and encrusting life, browsing on bryozoans, hydroids, even sea anemones. Some nudibranchs can even store intact, for their own protection, the stinging cells of the animals they prey upon. Nudibranchs lack gills, and thus use their body surface or projecting fleshy horns, called cerata, as respiratory organs.

down with a cold. When I finished I swore I'd never dive again in cold water." Vows are often forgotten, however, and eventually—off Alaska's mist-shrouded islands—Faulkner learned the secret to exploring frigid seas in relative comfort: properly thick rubber garb, of course, plus a large thermos of hot water, to be poured a little at a time into gloves and wet suit.

Amchitka, Baranof, Vancouver, Santa Catalina—the marine environments of all four islands are influenced by the Oyashio, a cold ocean current that flows south out of the Bering Sea and is so laden with microscopic animal and plant life that it is recognizable across vast expanses of sea by its green color. (Warm currents, with a dearth of planktonic life, are a transparent blue.) Off the coast of Japan, the Oyashio adds its wealth to the North Pacific's great clockwise gyre of currents that bathes the western shore of North America. And it is this plankton that sustains, directly and indirectly, so much of the life in these cold waters.

That life, to the unenlightened eye, may seem quite unexciting and unremarkable. In contrast to the airy azure realm of the coral reef, cold-water habitats are dark and somber. The bottoms of northern seas are not extravagant murals of living creatures, but instead are sediment-laden and punctuated only by muted patches of color. It is the camera's inquisitive lens and instantaneous flash that reveal the truth. There is wondrous life here, like the bizarre ratfish and wolf eel, that cannot be found anywhere in tropical waters. There are fish like the garibaldi that need not defer to their variegated tropical relatives. Somehow, cosmopolitan creatures like sea stars, sea anemones, and nudibranchs seem more brilliantly colored in colder seas than others in the reef world. And their behavior may differ sharply: sea pens, for example, feed at night on the reef, but are active during the day off northern shores. Always, for the photographer and for the reader, there is a sense of discovery around every submerged rock wall, on every turned page.

A sea urchin has wondrous armament; if a passing fish casts a shadow on its body,
the needle-sharp spines—which may be a foot long and in some species are hollow
and filled with poison—quickly turn in the direction of the potential enemy.
Each spine, moreover, has a ball-and-socket joint under individual muscular control.
The joints are the symmetrical rows of bumps that decorate the thin and lovely shells.
Sea urchins have five sharp teeth around a circular mouth, and they grind away
on a wide variety of foods, such as a jellyfish that has become impaled on the crimson spines.

*For its nest site, the boreal chickadee
with its brown cap chose a rotting stump.*

The Birds of
Summer Island

PHOTOGRAPHS AND TEXT BY ELIOT PORTER

GREAT SPRUCE HEAD ISLAND *is a molar tooth in the middle
of Penobscot Bay, the largest bay on the coast of Maine.
It is three hundred and twenty acres of gravelly beaches, tidal inlets,
steep rocky headlands, dripping fern-draped cliffs,
mossy outcrops, spruce forests, sphagnum bogs,
mucky swamps, and meadows. In 1912, when Eliot Porter
was ten years old, his father bought Great Spruce Head
as a place to take his family in summer. It became, then,
Summer Island, and its wildness and wild inhabitants
have lured Eliot Porter to its shores for sixty years.
For nearly four decades, having abruptly changed
his life's goal, he has gone there with camera
to record its special beauty. And its special birdlife.*

114

A pair of red-eyed vireos at their sturdy cup nest,
built with weeds and the pliable fibers of weathered bark,
held together with the silk from spider webs and caterpillars,
and suspended by its rim from the fork of a twig.

ly hummingbirds are smaller than the winter wren,
th its skimpy, tilted, reddish tail. And few nests
more skillfully hidden. This structure of pine twigs
s concealed within a curling sheet of birch bark.

117

*Only the male ruby-crowned kinglet has a ruby crown,
and the crown is seldom seen in the field. Its nest
is a bulky cup woven from mosses and lined with feathers
and is hung from drooping twigs high in a spruce.*

*. male Blackburnian warbler, a species that nests
.nd forages in the topmost branches of spruce trees.*

The white-throated sparrow's nest is sunk into a mossy hummock
decorated with starflower and wild lily-of-the-valley.

Eliot Porter on Bird Photography

❡ My interest in birds goes back to my childhood and the influence of parents who encouraged a preoccupation with Nature. And for reasons I cannot satisfactorily explain, capturing birds on film in spontaneous positions and in their natural surroundings seemed a most wonderful pursuit. Under different circumstances, I might have become a wildlife painter. But the camera could produce faster results than pencil or brush.

❡ My love of birds developed steadily along lines not usually recognized as justification for a full-time profession. Photography in those early years was my way of establishing illusory rapport with the secret lives of wholly unapproachable creatures. But I have always been a great deal more affected by the beauty of birds than by the mysteries and unanswered questions concerning their behavior and classification that occupy the ornithologists.

❡ My first groping photographic efforts were exclusively with birds. Over the years my work became more general until—with improvements in skill and advances in technology—I returned to birds. I soon realized the criteria of excellence applied to bird photography were considerably below the high professional standards required of photographers in other fields.

❡ A new point of view was necessary to raise bird photography above the level of mere reportage. Simply recording a bird's image on film was not enough. The entire picture area was vitally important. Every element in it must contribute to the unity of the image if the picture is to merit consideraton as art.

❡ The best time to photograph birds is while they are rearing their young. Of course, the nest— which will be the center of the pair's predictable activites for several weeks—must first be found. This is the bird photographer's most time-consuming chore. It is also the most enjoyable, for it keeps him out in the wild places for hours at a time.

❡ To find the nests of particular species, the photographer must be skilled at identifying birds by sight and song. He must know the habitats they require, the nest sites they prefer, the material used in the nests and how they are put together. But first he must be familiar with the geographic distribution of birds, their climatic preferences. It would be folly to look for birds of the forest in the meadow, or desert species in New England.

(The number of nests found is directly proportional to the time spent searching. But discoveries occur when least expected, and the bird photographer must be constantly alert. Frequently I have concentrated on locating the nest of one particular bird—and quite unexpectedly come upon the nest of another species.

(The bird photographer should not count his pictures before he has taken them. Nor can he afford complacency on finding a single nest. Many hurdles remain. If the nest is unduly disturbed during construction or incubation, it may be deserted. And any one of a number of predators—jays, grackles, shrews, weasels, chipmunks, squirrels, snakes—may destroy the eggs or young before a single frame is exposed.

(Most birds are extraordinarily adaptable. Once they have accepted the presence of a camera close to their nest, they scarcely notice the addition of a formidable array of equipment—tripods, flash lamps, humming power packs, electronic triggering devices—or even the photographer himself. Warblers are particularly fearless. To photograph a cerulean warbler nest high in the trees, I had to build a forty-foot tower. The birds accepted the intrusion with such indifference that I could stroke the female as she brooded her young.

(The bird photographer bears a great responsibility, for his activities must not jeopardize the successful rearing of the young. If they are not being fed frequently enough, or are exposed too long to sun, rain, or cold, he should withdraw immediately. If he returns to the nest later, it must be with the utmost caution.

(The warblers, because of their beauty and varied nesting habits, hold for me a special interest. But sparrows and flycatchers challenge the photographer to portray precisely their subtle distinctions of markings and color.

(Bird photography, if seriously undertaken, soon involves one in projects that a beginner could hardly anticipate.

The Jury Is Still Out

PHOTOGRAPHY BY CAULION SINGLETARY

TEXT BY FRANK GRAHAM JR.

THE ISLAND lay a few yards across the narrow waterway from the mainland, but only a dim shape, lacking substance, marked its presence in the fog. Joe Jacobs and I stepped into the small boat and prepared to cross to the island. And then, from somewhere in the murk ahead, rose the high whistled *whew! whew! whew! whew!* of the bird we had come to find.

Joe Jacobs has been banding ospreys on New Jersey's southern coast for nearly thirty years. Few men know that coast and its ospreys as well as he does. What he has seen of both in recent years sometimes turns a labor of love into a sad and frustrating experience.

It was an early morning in late June, and a rising sun was still trying to burn the fog off the marshes as we landed on Cedar Island. Looking back to the mainland, we could barely discern the row of buildings that make up the dense clutter of the once sleepy little borough of Avalon. Laughing gulls yelped somewhere around us, and a glossy ibis sliced through the fog and disappeared on some errand.

Breeze and sun were shredding the fog now. We trudged through the gulping mud that lay beneath the expanse of marsh grass. A couple of willets, large sandpipers whose glory is revealed only when they open their flashing black-and-white wings, flew excitedly around us, uttering the high-pitched cry that gives them their name.

"The willets began nesting again along the Jersey coast about twenty years ago," Jacobs said. "The gunners and egg collectors had driven them out in the last century. I still haven't found one of their nests on this island; but from the way these birds are behaving, I'd say we're pretty close to one."

It was apparent that we were coming close to an osprey nest, too. We could see two adults now, flapping above the marsh in narrowing circles, their cries growing sharper and more frequent as we approached. Two huge nests were just ahead of us, both built in dead red cedars.

"The one in the tree to the left doesn't have anything in it," Jacobs said. "I checked it out a

124

few weeks ago. But there were eggs in this one."

We stopped in front of the cedar on which Jacobs was pinning his hopes. The nest sat in the top of the bare tree, about fifteen feet above the ground. It was built mostly of sticks and the marsh grass *Spartina alterniflora*.

Jacobs, a slender, graying man in his mid-fifties, began to ascend the tree under a crescendo of cries from the circling parent ospreys. An English sparrow flew out of a hole in the bottom of the ospreys' nest where it had found a convenient niche for a nest of its own. Jacobs climbed spryly and confidently.

"These cedars are great trees to climb," he said, grinning down at me. "There's no rot in them for years after they've died. Even these thin branches will hold a man—or a nest."

Jacobs pulled himself up to eye level with the nest and looked over the rim. Then he looked down at me, nodding with satisfaction.

"There's one young bird in here," he said.

He reached in and lifted the young osprey for my inspection. It was a forlorn-looking little thing, its neck and heavy wings drooping as it feigned death, a state belied by its bright orange eyes. Feathers, buffy and darker, grew among its natal down, and its black-and-white head pattern had emerged. The bird was about five weeks old.

There was a distinct *whoosh!* as a parent osprey swooped to within two or three feet of Jacobs' head. He set the nestling down and began fastening a large aluminum band to its left leg.

"Anything else in there?" I asked.

He shook his head. "Nothing but the usual junk you always find in an osprey nest—a board that's painted red, a length of fishing line, a chocolate drink carton. They're real collectors. Once I found a dead muskrat in a nest, not eaten or pecked at, but just there for decoration. Some people have found clapper rails. The ospreys hadn't killed them; they just found their carcasses on the marsh and brought them in to give bulk to the nest."

Jacobs finished his work quickly and descended.

"I don't believe in disturbing these birds any more than necessary," he said as he stuffed his notebooks and extra bands into a canvas bag. "Some of these researchers are going to wear out the birds they're working on."

His banding of osprey nestlings has a long tradition on Seven Mile Beach and the other salt marshes along the southern New Jersey coast. John Gillespie of the Delaware Valley Ornithological Club started banding young birds in the nest in that area in 1926, and others, chiefly Jacobs, have carried on for him. Ninety-three bands have been recovered. Just before we set out on our expedition, Jacobs learned that an osprey he had banded as a nestling in 1956 had been recovered early this March in Camaguey, Cuba.

In fact, much of the information that is helping to clear up the confusing picture of osprey populations may be attributed to the work of nonprofessionals. Joe Jacobs is an amateur in the truest sense of the word. He is a carpenter by trade—a painstaking craftsman in an era of shoddy workmanship. But his love for birds, and especially ospreys, leads him into the field at every spare moment. He not only bands the ospreys, but works actively to protect them and to encourage their nesting. And he has helped document the crash of the osprey that first puzzled and then alarmed ornithologists.

Jacobs spoke about his efforts as we set off again through the lifting fog on Cedar Island. The cries of the parent ospreys had subsided behind us, but willets still fluttered overhead, and the harsh *kek-kek-kek* of clapper rails sounded from somewhere in the dense spartina. We hardly noticed the increasing clatter of boat engines in the waterways around the island.

"I got interested in birds when I was a boy," he said. "I grew up outside of Camden, but I began to spend my summers down here on the coast. It was a marvelous place for anyone who loved birds. Why, when I built the cottage here we had yellow-crowned night herons nesting in the backyard, and ospreys nesting down the street."

Avalon has changed greatly since then. It is a dense gridwork of unshaded summer cottages, stores, and hamburger joints. Bridges crisscross the once uninterrupted stretch of marshes. Beach buggies and trail bikes rumble over the dunes and sandy stretches where skimmers, terns, and plovers once nested.

"Thirty years ago I could go out in a rowboat and band forty young ospreys in a single day," Jacobs recalled. "They nested everywhere—on telegraph poles and fishermen's shacks and duck blinds. Now I go out by car and motorboat, and I'm fortunate to find ten young birds in a day."

Year by year more nests were abandoned, or the eggs they produced did not hatch. At first Jacobs believed he might be able to reverse the trend by providing the birds with nesting places isolated from man's encroachment.

"I put up my first nesting platform right here on Cedar Island," he said, indicating the marsh that stretched around us. The fog had lifted now, unveiling the nearly square island of about two hundred and ten acres. On the east side of the island, clumps of cedars grew on the remnants of ancient dunes. The lifting fog also revealed the teeming clutter of the summer resort at Avalon just across the waterway.

"There used to be sixteen pairs of ospreys on Cedar Island, and they were doing pretty well," Jacobs said. "The state offered to buy the island as a refuge, but the borough of Avalon held them up. Avalon wanted to put a bridge across to the island and develop it with homes and marinas. A few years ago the state finally had to use $135,000 in Green Acres funds to buy it."

But Jacobs noted that there were not enough good nesting sites for all the ospreys that seemed to want to nest on Cedar Island. He decided to set rectangular platforms on stakes about four feet above the marsh.

"You probably couldn't do this in an area where the ospreys were used to nesting high in trees or on cliffs," he said. "But the ospreys around here had often nested low—sometimes they even built their nests right on the ground around an old stake, or on a derelict boat.

"In 1967 I came out and hammered together the first platform from scrap wood and turkey wire after the birds had already arrived from the south. Then I picked up some spartina and a few sticks and tossed them on the platform—just to give the birds the idea. When I got to the boat and looked back, two birds were already on the platform."

Jacobs built several dozen nesting platforms in the marshes around Avalon. When winter's ice carried one away, he replaced it the next spring. The ospreys occupied most of the platforms immediately.

"There was an old nest used year after year next to a waterway that's heavily traveled by pleasure boats," Jacobs said. "The nest never produced any young. So in 1970 I put up a platform a couple of hundred yards into the marsh away from the thoroughfare. The birds abandoned the old nest and built a new one on the platform. In 1971 they raised two young, in 1972

Using her powerful beak like scissors to snip pieces of flesh from a sheepshead, the female osprey feeds her thirty-nine-day-old nestling. The bottoms of her feet are adapted for holding slippery fish, and that well-placed stick is used as a brace during the three or four daily feedings. There are two young, and each will be fed equal portions until they are full. The rest of the fish—including the tail—will be eaten by the mother. At this age, she spends little time on the nest, but perches nearby on a mangrove stub.

they raised one, and this year there are two young birds in there again."

Despite Jacobs' untiring efforts, the number of young ospreys fledged in the area has declined in recent years. In the thirty-odd nests around Avalon and Stone Harbor observed by Jacobs this year, he found only eleven young. As we tramped across the marsh on Cedar Island I began to understand his frustration.

"The birds were incubating eggs in this nest a month ago," Jacobs said as we came to one of his platforms. It was the last of the eleven "active" nests we had visited in trees or on platforms on Cedar Island, and we had found but a single young bird.

There were no traces of eggs or young in this nest either. There was the jaw of a hammerhead shark, probably picked up by an adult bird on the beach for decoration. There was a plastic baseball bat and a length of plastic garden hose woven among the sticks and grass. Rising from one corner of the big nest was a flourishing sprig of lamb's-quarters, the only life the nest would produce this year.

Joe Jacobs' experience on the marshes around Avalon, New Jersey, is but one element in the complex story of what is happening to one of our most magnificent birds of prey. The story is not at all consistent. Several years ago a trend was apparent along the coast of the northeastern United States, and it was thoroughly disheartening. Now the jury is out once more.

The osprey, or fish hawk, frequents both fresh and salt water areas around the world. It is a single species, divided into five subspecies, with the North American race known as *Pandion haliaëtus carolinensis*. It inhabits even remote islands where other birds of prey cannot sustain themselves. Like so many birds, it takes its name from an early human ignorance of its characteristics: the word osprey derives from the Latin *ossifragus*, meaning "bone-breaking." Yet this bird feeds almost exclusively on fish.

The osprey is neither a scavenger nor pirate, as is the bald eagle. Its spectacular dives for live fish have delighted everyone who takes an interest in watching birds and other wild creatures. Circling high in the air, sometimes up to two hundred feet, an osprey detects its prey, hovers, and then plunges toward the water with its wings half closed and its feet extended. For a moment it disappears below the surface in a burst of spray. Then, flapping laboriously, it lifts itself out of the water with a fish firmly grasped in its inch-long talons.

Unlike most birds of prey, ospreys and their huge nests (perhaps five feet across) have never been strangers to human beings. They have often nested around buildings and farmyards. The species was driven from the British Isles at the beginning of this century by gamekeepers and egg collectors; but in America, at least until recently, it managed to survive and even flourish. Some farmers regarded ospreys as a sign of fortune and encouraged their presence, as Europeans encourage nesting storks, by erecting wagon wheels as nest supports on poles and trees.

Ospreys often breed in colonies, with dozens of nests built within sight of each other. The decline of some of these colonies is not a phenomenon peculiar to our own time. Large colonies dwindled in southern Massachusetts in the last century, a fact attributed by some recent experts to the gross pollution of local rivers and estuaries by the wastes discharged by industries.

The great colony (estimated at over two hundred and fifty nests) on Plum Island, at the east end of Long Island Sound, disappeared when a summer resort developed there; but the birds moved almost en masse to nearby Gardiners Island and built one of the largest osprey colonies ever recorded—perhaps three hundred pairs of birds.

Egg collectors were blamed for the decline in osprey numbers along New Jersey's Seven Mile Beach in the late nineteenth century. One oologist took the eggs from thirteen nests in a single day. And ornithologist A. H. Norton explained the decline of several Maine colonies this way: "Persons of mature and well-ordered minds seldom molest these birds, which are regarded as harmless and industrious. There is a degree of reverence, and even superstition, against killing an osprey on the Maine coast. During the two decades of 1880 and 1890, there was an influx of foreign-born people into the state, whose primitive conception of hunting was abetted by an abundance of cheap fowling pieces and ammunition. Reverence and superstition were ignored. These fine birds paid the penalty of being abundant and solicitous of the welfare of their nests; the birds were attractive targets for the blood-thirsty beings who deemed themselves 'sportsmen' in destroying hawks."

Milliners perverted the word osprey for their own purposes during the height of the plume craze in the nineteenth century. When outrage mounted against the slaughter of herons and egrets for their plumes, public-relations-oriented milliners insisted on calling the plumes ospreys instead of aigrettes. Perhaps the idea was that softhearted ladies would be mollified if they thought the plumes were taken from some murderous bird of prey rather than from a gentle egret. Fortunately, ospreys do not produce the plumes coveted by milliners.

On the whole the big birds stabilized their populations once Audubon societies and other conservation groups secured minimum protective measures for them. They graced whatever landscape they inhabited. In his *Bird Studies at Old Cape May*, Witmer Stone defined the osprey's place in his part of the country:

"I always think of the fish hawk as the connecting link between the life of the sea and of the upland and the first intimation of a change of environment as we travel through the pinelands and scattered farms of South Jersey on our way to the coast. The immense nest topping some old sour gum tree along the fence row, the great birds circling overhead bearing shimmering fish in their talons, their fearless solicitude for nest and young, their apparent confidence in the security of their exposed eyrie and the querulous whistling of both old and young—all make up a never to be forgotten picture of that delectable borderland where the elements of land and sea meet and intermingle."

Roger Tory Peterson has written that he moved to Old Lyme, Connecticut, in 1954 largely because of the colony of ospreys at the mouth of the Connecticut River. "There were about 150 nests in the general area, and they were a pleasure to everyone," Peterson wrote.

But not all was well with American ospreys. Ancient prejudices against birds of prey persisted. According to Frederick C. Schmid of the U.S. Fish and Wildlife Service, eighteen ospreys were shot at a private fish hatchery in 1964 in a state where these birds were protected by law. The

131

owner of the hatchery justified the shooting with the comment that he was not raising trout for ospreys to eat.

Ospreys were shot at state hatcheries, including those in Utah, during the 1950s. They were also shot at a number of federal fish hatcheries until John S. Gottschalk put an end to the practice during his tenure as director of the Bureau of Sports Fisheries and Wildlife in the 1960s.

In the Gulf of California, only those ospreys nesting high on cliffs survived. Observed Lewis Wayne Walker: "I find that it is the low nests that are no longer occupied—the nests that are easy targets for a .22-caliber rifle from a rocking boat."

In the Chesapeake Bay Area the prime villain was the United States Coast Guard. According to Jan G. Reese, who has studied ospreys in the bay for some years, coastguardsmen routinely destroyed any nest, including eggs or young, that they found built on buoys and other navigational aids. Reese says that despite repeated protests by conservationists, the Coast Guard destroyed 9 percent of all the nests studied from 1965 through 1970. Only in recent years have the protests fallen on receptive ears.

But early in the 1960s it was apparent that there was a decline in ospreys that could not be accounted for simply by the depredations of trigger-happy sportsmen and hatchery operators, or even by runaway land development. Many osprey colonies dwindled regardless of their isolation or degree of protection. The tremendous colony on privately owned Gardiners Island, isolated and absolutely protected against incursions by human beings and predatory mammals, was decimated almost as completely as the colonies along the booming South Jersey coast.

By 1969 the Gardiners Island colony held only thirty-eight active nests. The Connecticut River colonies, despite the care and concern of Roger and Barbara Peterson and their friends, crashed from seventy-one nests in 1960 to seven in 1967. In the same period (1961–69) Rhode Island's active osprey nests dropped from sixty-one to seven. Paul Spitzer of the Cornell Laboratory of Ornithology, drawing on old records, estimated that there were eleven hundred active nests between New York and Boston in 1940. By 1969 he could find only one hundred and thirty-five.

The crash was not confined to the Northeast. Although there were significant exceptions to the trend, notably in Chesapeake and Florida bays, reports from many parts of the country indicated that the big birds were in trouble. One of the most impressive studies was made by Sergej Postupalsky of the Detroit Audubon Society during the middle 1960s.

"I know of twenty-five [osprey] territories in Michigan which were abandoned between 1964 and 1967," Postupalsky wrote. "This amounts to 27 percent of all known nest sites in that state."

Even more disturbing was his discovery of the poor reproduction in those nests that were occupied. Surveys made before World War II in several Eastern states indicated that "normal" osprey colonies produced about 1.6 young birds for every nesting pair (including those that were unsuccessful in hatching young). Michigan birds produced .36 young for each nesting pair in 1965. This figure dropped to .30 the next year.

"Nest success is equally low in both peninsulas—in the highly developed Lower as well as the less developed Upper Peninsula," Postupalsky wrote. "Also my data indicated that there was no significant difference in nest success between the accessible or exposed sites and the more isolated wilderness sites. . . . This lower production is due to high egg losses or the failure of eggs to hatch."

This phenomenon was noted over and over again. In northern Wisconsin's Rainbow Flowage, twelve of twenty-seven nests were successful in 1955, producing twenty-two young birds. Ten years later only seven nests remained, and one young osprey was raised in the entire colony. Peter L. Ames of Yale University, working with the Petersons in the Connecticut River estuary, found that less than 20 percent of osprey eggs (and sometimes only 5 percent) hatched. Many contained dead embryos. Even when the researchers erected special raccoon-proof nesting poles, they found broken eggs in the nests.

That the osprey, like most other wild creatures, was under heavy pressure from human development schemes seemed undeniable. Although it can tolerate a certain degree of proximity to man, the osprey obviously cannot survive in the midst of urban and suburban sprawl. But evidence was piling up that suggested the chief reason for this raptor's low rate of reproduction lay within itself or its egg.

The osprey is especially vulnerable to long-lasting contaminants that are passed up through food chains. Standing at the top of its own food chain, it takes in residues that were concentrated as they moved from one trophic level to the next. Now suspicions about DDT's insidious hazards were being confirmed. Ornithologists such as D. A. Ratcliffe in Great Britain and Joseph J. Hickey in the United States were showing that the thickness of some birds' eggshells, when compared to those laid in the pre-World War II (and pre-DDT) era, was considerably diminished. A sample of twenty osprey eggshells collected in Connecticut, for instance, measured 25 percent thinner than the average shells collected before the introduction of DDT.

Laboratory experiments confirmed the relationship, showing that some birds fed DDT laid eggs with thinner shells than the controls did. "When one picks up an egg that has been thinned by pesticidal doses," said William Stickel of the U.S. Bureau of Sport Fisheries and Wildlife, "the egg is cracked by the touch of the fingers."

Ospreys, then, were laying some eggs that cracked easily in the nest. This explained the increasingly frequent reports of eggs disappearing from the nest early in the incubation period. They usually disappeared one at a time, suggesting that they occasionally broke and were tossed out of the nest by the parent birds. If predators had been responsible, all the eggs would have disappeared simultaneously.

Research has proved that the thickness of osprey eggshells is usually related inversely to the concentration of DDE (DDT's breakdown product) in the eggs. Cornell's Paul Spitzer has summarized some of the latest work:

"The reduction in eggshell thickness in eggs from Connecticut, particularly the discovery that eggs found damaged in nests early in the incubation period were relatively thin-shelled, and the finding that no Connecticut eggshell in the recent sample was thicker than the pre-1947 norm, indicate that eggshell thinning has probably been an important factor in the population decline and reduced reproductive success of Connecticut ospreys. . . . In a variety of bird species where excessive eggshell thinning (15 percent or more) [had] occurred for a period of years, the population involved was in trouble."

Spitzer and others carried out the now-famous exchange of osprey eggs between successful Maryland colonies and unsuccessful Connecticut nests. Banding returns had already indicated that most of the ospreys that breed in the Northeast spend the winter south of the United States. Moreover, ospreys from the New York, New Jersey, and Maryland areas had the same

Its eyes closed in apparent concentration or pleasure, a forty-six-day-old osprey carefully preens the feathers on its back. Each feather is tipped in white as though dipped in paint. At a warning cry from a parent, the young bird flattens itself against the nest, its mottled plumage excellent camouflage.

136

Now fifty-three days old, its first flight only two days away, the young osprey
exercises its wings in powerful beats that can lift it several inches
above the platform. The fledglings will continue to be fed at their nest,
and they will use it as their favorite perch as training continues.

general distribution over their common wintering areas, and so it was likely that the breeding success of any colony depended upon the environmental conditions on their breeding grounds.

Beginning in 1968, freshly laid eggs were taken from Maryland nests that had had a history of successful production (40 to 50 percent) and placed in Connecticut nests, which had had a low production rate of 5 to 15 percent. Conversely, Connecticut eggs were flown to Maryland. Connecticut eggs held higher concentrations of DDT, DDE, dieldrin, and PCBs than did Maryland eggs, and on the average their shells were thinner. Connecticut fish were more contaminated. In conclusion, the researchers had this to say:

"Incubation of Connecticut osprey eggs by Maryland ospreys did not result in an improvement in hatching rate. Maryland osprey eggs incubated by Connecticut ospreys hatched at their normal rate. The results of the egg exchanges and associated observations indicated that the most probable cause of the poor reproduction of Connecticut ospreys was something within the birds and/or their eggs. . . . Three environmental pollutants that have most likely been important factors in the greatly reduced reproductive success and rapid population decline of Connecticut ospreys are dieldrin, DDE, and PCB."

The connection is not always clear, and the facts are sometimes inconsistent. There is not always a correlation between DDE levels and eggshell thinning. Some birds laying thin-shelled eggs produce young, while embryos die inside eggs with thicker shells. DDT, DDE, and dieldrin are more prevalent in Connecticut eggs, while PCBs predominate in those laid in some other regions.

It would be foolish to put all the blame on a single substance. But despite the sort of variability that frustrates statisticians, the mass of evidence has established the crucial role of these contaminants in the crash of osprey populations. Another fact is clear: during the last decade many osprey colonies in North America have been reproducing themselves at only a fraction of the rate that is necessary for their survival.

It was on the basis of such information that Joseph J. Hickey, now president of the American Ornithologists' Union, predicted in 1969 that the osprey soon would be wiped out from Maine to New Jersey.

What is the osprey's present status in this country? The federal Committee on Rare and Endangered Wildlife Species lists the osprey as "status undetermined," and that sums up the current picture. In some areas its prospects seem poor, while in others there is reason for hope. New factors are emerging.

The large colony in Florida Bay, though fluctuating occasionally, remains stable at about one hundred and fifty nests, and may even be increasing. A recent survey by the Bureau of Sport Fisheries and Wildlife found nine hundred and seventy-five active nests in the Chesapeake Bay Area (one of them built on a practice bomb that had been dropped on a bombing range).

But a prediction made ten years ago by Roger Tory Peterson has been fulfilled: osprey production in the Connecticut River estuary has hit zero. To the hazard of infertile eggs has recently been added the rapid disappearance of adults. Age is taking its toll, and perhaps other factors as well—a mature Connecticut osprey was found dead, its brain contaminated by dieldrin residues. Woolen mills on the river use large amounts of dieldrin in mothproofing.

The federal ban on DDT has given many ornithologists cause for hope. When dieldrin was banned from sheep dips in Scotland, residues of that chlorinated hydrocarbon pesticide in the eggs of golden eagles decreased significantly, and the proportion of golden eagles raising young successfully more than doubled. Ospreys also have returned to Scotland and are reportedly doing well. And areas in this country where high DDT residues had appeared in osprey eggs began to show a better record of fertility as soon as DDT was phased out.

"Eastern Long Island used to be saturated with DDT," says Richard L. Plunkett of the National Audubon Society. "As the insects became increasingly resistant to it, the potato farmers increased the dosage, using up to thirty times the applications recommended by the U.S. Department of Agriculture. Even that didn't work for long, and so DDT was phased out in favor of more effective, shorter-lived pesticides in the mid-1960s. A lot of DDT was used in mosquito control, too, but that was halted by a lawsuit in 1966."

As DDT was phased out on eastern Long Island, a higher percentage of osprey eggs began to hatch. Osprey reproduction had reached its lowest point on Gardiners Island in 1965 and 1966, when only four young were fledged from fifty-five to sixty nests each year. Since then the fledgling rate has risen, and in 1973 there were eighteen young birds produced in thirty-one active nests (in addition to four hatched from introduced Virginia eggs).

But on Gardiners Island, as in Connecticut, new problems have emerged.

"We're getting more young birds, but there are fewer active nests every year," says Dennis Puleston, who has observed the Gardiners Island colony since 1948. "The old birds are dying off, and they aren't being replaced. And now that DDT is gone, the eggs are hatching all right, but the fledglings are dying."

Puleston is convinced there is now a food problem.

"The commercial fisheries are in trouble," he says, "and the ospreys aren't having any luck either. They have to go farther every year for their fish. There's been a big drop in menhaden, for instance. That used to be the fish the ospreys usually fed to their young. Now the ospreys are subsisting on sea robins and small flounder, bottom fish which are harder to catch. When we band the young birds, we find they're starving."

Herbert H. Mills, the late conservationist, was one of the first to mention the osprey's diminishing food supply.

"Since 1966 we have given particular attention to the remaining birds in southern New Jersey," Mills told the North American Osprey Conference at Williamsburg, Virginia, in 1972. "We have watched as many as five or six birds fishing in the ocean gullies in front of our house day after day, and have seen them returning after many, many attempts with empty talons or with fish so small that one wonders how they were able to hold them."

Mills, shortly before his untimely death in October 1972, experimented by leaving menhaden on well-used perches near osprey nests. It was an offer the ospreys could not refuse, and they took the fish readily.

Indeed, man, who is to blame for the osprey's present plight, also seems to be playing an important part in maintaining whatever populations are left. The construction of predator-proof nesting platforms in many areas has helped. The U.S. Forest Service is cooperating actively with the National Audubon Society in making surveys of osprey colonies on its lands and in setting aside management areas during the breeding season.

In the Far West, successful osprey colonies have formed around man-made impoundments and reservoirs. One of the most notable is that at the Crane Prairie Reservoir in Oregon's

Deschutes National Forest. The reservoir is the center of the largest concentration of osprey nests in the state, with the number of active nests increasing to sixty in recent years and producing an average of better than one young bird for each nest.

The exchange of Connecticut and Maryland eggs showed promising results. At least seven ospreys introduced as either eggs or young to Connecticut nests from Maryland three or four years ago have reappeared in Long Island Sound as adults. Three of them have nested.

The technique called double-clutching has been successful in Virginia. There, under Dr. Mitchell A. Byrd of the College of William and Mary, researchers collected the first clutch of eggs from nests known to have produced young consistently in the past. They put these eggs into consistently unsuccessful nests in Virginia and elsewhere, while the females in the original nests went on to lay a second clutch. Using this procedure, the researchers induced the ospreys to fledge more young birds than they had in previous years.

One of the most heartening stories comes from southern Massachusetts. The osprey colony at Westport has been in trouble for a number of years. Allen H. Morgan recorded the colony's decline for several years (and still recalls the day he was descending a tree when a rotten osprey egg exploded, with a dreadful bang and a worse odor, in his shirt pocket).

When Morgan's duties as executive vice-president of the Massachusetts Audubon Society forced him to give up regular osprey research, he turned over the Westport colony to a local couple, Gilbert and Josephine Fernandez. It was an ideal choice. Gil Fernandez, a retired teacher and a longtime Audubon member, had the time and background to do the job well. He and his wife live near the osprey colony on the Westport River, an estuary just east of the Rhode Island line. The big colony had diminished during Morgan's survey, and when the Fernandezes took it over eleven years ago it was down to twenty-two pairs of birds.

They threw themselves into the project with enthusiasm. They surveyed the birds' numbers. They erected nesting platforms and protected the ospreys against raccoons and humans. They banded the nestlings and did "public relations" work for the birds, giving lectures and slide shows. Nevertheless, the colony declined to fourteen pairs, and the production of young birds remained poor.

Despite the decline, there were several elements working in the ospreys' favor. About eight years ago Westport put strict limitations on the use of DDT. The osprey colony there is reasonably isolated from human disturbance. And the relatively clean estuary offers good fishing, its shallow water abounding in flatfish and striped bass.

Then, in 1973, there was a dramatic surprise. The fourteen pairs of ospreys all laid eggs, seven producing clutches of four. One pair fledged four young, and the colony produced a total of twenty-eight. The average of two young per nest rivals the success of healthy colonies before World War II.

"I don't want to say we've got a trend," Gil Fernandez says. "We had a pretty good season a few years ago, and then it dropped off again. But we're very pleased. We have new hope."

That's what Gil and Joe Fernandez, Joe Jacobs, and other people who are devoted to ospreys have been living on for a long time. Jacobs, who has seen no such heartening event among the South Jersey colonies, nodded with satisfaction upon hearing the news from Westport.

"I just feel optimistic," he said. "The bald eagle is gone as a nesting bird around here. So is the peregrine falcon. But these few ospreys keep coming back every year. They *persist.*"

140

Poetry in the Fields

PHOTOGRAPHS BY LES LINE

TEXT BY HAL BORLAND

I N THE book of Genesis we are told that Adam named "every living creature." Nothing is said about naming plants, but for a long time they had only the kind of names Adam might have given them, or more likely, Eve. Simple, descriptive, poetically picturesque names such as blue-eyed grass, ladies'-tresses, adder's-mouth, lady's-slipper. But eventually scholars tried to list the plants and found a confusion of names from place to place. Finally Theophrastus, a student of Aristotle, tried his hand at it and compiled an herbal, a plant list, based on the most commonly used names. When the Romans succeeded the Greeks, they tried to improve the herbal by making the names descriptive. Eventually they had the names so long that even the scholars couldn't understand them.

Meanwhile, the ordinary people went on using the common names. So did the herb dealers and herb doctors, who had more practical knowledge about plants than anyone else. This duality of names continued until 1735, when Carl Linnaeus, a young Swede, devised a simple, workable system of nomenclature based on Greek and Latin. Linnaeus gave us a precise international plant language that can be understood anywhere. But the old names persisted, and still do. Webster's unabridged dictionary lists more than 150 "wort" names for plants still in use. Wort comes from the Old English "wyrt" and means plant or root. The herb folk simply tacked a descriptive word onto it—feverwort, for example, sometimes called horse gentian, and botanically *Triosteum perfoliatum,* was a plant from which the herb doctors brewed an infusion to reduce fevers. Sneezewort was supposed to stop sneezing, cankerwort was good to treat cankers, and fleawort may have helped rid a dog or a person of fleas.

So we have this dual system, which overlaps at many points. The scientific names are specific. The common names are sometimes local or regional and may vary from place to place. They persist because they are a part of the common language, because they are often picturesque, and because they have folk poetry in them.

Continued on page 153

142

Our familiar day's eye, the oxeye;
stargrass, demure kin of the daffodil;
and hawkweeds in sun-orange and gold
to sharpen the eyesight of birds of prey.

From the rainbow, our lovely wild iris or flag;
its unassuming relative, blue-eyed grass;
from both doves and eagles, the crimson columbine;
and Deptford pink, "small eye" to the early Dutch.

146

In a maze of weeds, a wand of sour sorrel or dock; in a fall field of goldenrod, a bouquet of New England asters.

A wild lily named for a Turk's cap.

The rose of our swamps, Rosa palustris.

Let's go on with the worts for a moment. There is adderwort, a *Polygonum*, supposed to be good for snakebite. There is a whole group of snakeroots, one of which, *Aristolochia*, is also called birthwort and was said to be of value in aiding childbirth. There is nettlewort, actually any plant of the nettle family, and certainly no cure for nettle sting—I wonder if there is any cure? There's lousewort, *Pedicularis*, from the Latin word meaning louse, so named not because it kept lice away but because cows that ate it were supposed to get lice.

Common plantain was called ribwort, and the hedge nettle, *Stachys palustris*, was woundwort. A whole botanical family, Plumbaginaceae, is still called leadwort, though I can't find out why; the Latin name comes from the word meaning lead, the metal. At least two of the parsley family are common worts, honewort, also called wild chervil, and navelwort. *Dentaria* is known as toothwort, and perhaps was used to ease toothache or to help babies with their teething. *Achillea ptarmica*, one of the composites, is sneezewort, and the *ptarmica* is Greek for sneeze—you almost sneeze pronouncing it. Another composite, *Lapsana*, is called nipplewort and probably was used by nursing mothers.

You can go down the list interminably. But not all of them have medical meanings. Miterwort, a saxifrage, was so named because its seeds resemble a small bishop's cap, which is exactly what the botanical name, *Mitella*, means. Pennywort has leaves resembling small coins, and the botanical name, *Obolaria*, is from a small Greek coin, the obol. *Hepatica* is liverwort, so named because the leaves are liver-shaped. We find that botanical name also in hepatitis, a disease of the liver. Swallowwort is *Chelidonium*, the Greek name for swallow. We also call this plant celandine, and the old Greeks believed that mother swallows used its yellowish juice to wash out the eyes of their nestlings and give them good eyesight.

The overlap between botanical names and common names is often fascinating. Take *Scrophularia*, figwort. It comes from the Latin for scrofula, which it was supposed to heal. It was also supposed to cure fig warts, whatever they are. *Chelone*, the turtlehead, comes directly from the Greek *chelone*, meaning tortoise. *Dracocephalum*, dragonhead, is from the Greek *dracon*, a dragon, plus *cephale*, head. One of the most surprising is *Parthenocissus*, Virginia creeper. Greek *parthenos* means virgin and *cissos* means ivy—the virgin's ivy, or creeper.

Tragopogon, commonly called goat's-beard, is a literal translation to the Greek—*tragos*, goat, and *pogon*, beard. Burdock is *Arctium*, from the Greek for bear, and refers to the plant's bristly burr. Burdock is one of the composites, of course. But the docks themselves belong to the buckwheat family, and many are also known as sorrels, from the Old French, *surele*, meaning sour. Dock comes from the Old English *docce*, a dark-colored plant. *Myosotis*, the pretty little forget-me-not with its sky-blue petals and yellow eye, comes from the Greek *myos*, mouse, and *ous*, ear, referring to the short, soft leaves. *Anemone* is generally thought to come straight from the Greek word for wind, as in anemometer, to give us simply the Greek name for windflower. But there is also a legend that says that the crimson anemone of the Far East sprang from the blood of Adonis. In Semitic, Adonis was called Na 'mān, and there are those who say anemone is a corruption of Na 'mān. This also seems a bit farfetched to me.

Deptford pink had me looking at maps before I learned that Deptford was once an administrative division of London, now part of Lewisham. The name pink may be short for *pink eye*, which refers to the shape of the fragrant little flower, and comes from the Dutch *pinck oog*, meaning small eye. If so, it would seem that the color pink was named for the flower, not vice

153

versa. The botanical name is *Dianthus*, from the Greek *dios*, the god Jupiter, and *anthos*, flower. In other words, Jove's own flower.

The blue flag, *Iris*, comes unchanged from the Greek word for rainbow, while *flagge* was the Middle English word for a rush or reed. Daisy is from *day's eye*, also out of Middle English, and refers to the fact some English daisies close their petals at night and open them the next morning.

The wild roses all get their names from the old Greek *rhodon*, which meant the flower we know as a rose, and which was *rosa* in Latin. Somewhere back in the mists it came from the Orient to the Greeks. There are many wild roses, their specific names usually descriptive. *Rosa palustris*, for example, is our native rose of the swamp or wet place. *Rosa blanda* is almost thornless, or bland. *Multiflora* is many-flowered. *Canina* is the dog rose, which refers to its mean, hooked prickles. And *Cinnamomea* is the cinnamon rose, so named because it has something of a cinnamon fragrance.

The asters obviously are named for the stars; the word's root also gave us astronomy. And there are several dozen species, some named for leaf shape, some for color, some for people. To my mind the most spectacular of all are the big purple and gold New England asters, whose Latin name is *Aster novae-angliae*. Anyone who can't figure that one out can go stand in the corner.

The whole *Chenopodium*, or goosefoot, genus is named for the Greek *chen*, goose, and *pous*, foot; and it was called goosefoot in the first place because the leaves look very much like a goose's foot. The common pink woodland lady's-slipper is *Cypripedium*, from *Cypris*, the Greek name for Venus, and *pedilon*, shoe.

There are bafflers, of course. Remember the one about the anemone? The columbine, *Aquilegia*, is another. Some say the botanical name comes from the resemblance of the flower's spurs to the claws of an eagle, Latin *aquila*. Others prefer to believe it comes from *aqua legere*, to collect water, referring to the dew that collects in those saclike spurs. The common name apparently comes from an old Latin and Norman name for the dove, *columba*. It once was known as five crimson doves, the spurs thought to resemble doves. There is the large crowfoot genus, *Ranunculus*, from the Latin word for a little frog. The frog connection dates to Pliny, who noted that many of these plants were aquatic species that grow where frogs abound.

Another of the old legendary tales is recalled in the genus *Silene*, the campions. Silenus, foster-father of Bacchus, seems also to have been quite a toper and was covered with beery foam from time to time. Some of the campions have viscous excretions that look vaguely foamy. Hence one common name, catchfly. Bladder campion's source is obvious, and its fat orbs are likened to tears by some, who call the plant maiden's tears. Starry campion is also known as widow's frill for its fringe of white petals. And hawkweed, *Hieracium*, recalls another of the old superstitions—hawks once were thought to use its juice to sharpen their eyesight, a tale that recalls the one about celandine and the swallows. And one more old legend. Loosestrife, *Lysimachia*, from *lysis*, to free or let loose, and *mache*, strife. A legendary king of Thrace, chased by a bull, pulled up a plant of loosestrife, waved it at the bull, and the bull calmed right down.

Dogwood, *Cornus*, baffled me for a long time, but I finally ran it down. In the Middle Ages European butchers used the wood for skewers. In Old English the word *dagge* meant a dagger or sharp-pointed object. Hence, skewerwood or dagwood, corrupted into dogwood. And *Cornus*

154

is Latin for horn, referring to the hardness of the wood. I still haven't found out which Anne is honored in the naming of Queen Anne's lace, *Daucus carota*. And I am still a bit dazed by viper's bugloss, *Echium*. The botanical name is clear enough; it came from Dioscorides, who thought the plant's nutlets looked like a viper's head, and *echion* is Greek for viper. But then it gets murky. Bugloss, I find, comes from the Greek for cow and tongue. So what we come out with is a viper's cow's tongue.

Then there are those common names that sprang from other languages. Persimmon comes from the Algonquian word *pasiminan*, meaning dried fruit, or perhaps simply fruit. Squash is from the Algonquian also, *askutasquash*, the Indians' name for squash. Pumpkin, on the other hand, comes from the early English *pompion*. Brooklime, *Veronica*, is from Middle English, *broke* for brook and *lemeke*, a kind of plant. A waterside plant, in other words. Teasel, *Dipsacus*, is from the Old English *taesel*, to pluck or tease. Dandelion, of course, is right out of Old French, *dent-de-lion*, referring to the toothed leaves. But there was a slipup in the scientific naming, for instead of going to the common dandelion, the lion's tooth name went to a lesser plant, the hawkbit or fall dandelion, the one known botanically as *Leontodon*.

One of the most unusual namings is modern. A weedy field and garden pest with tiny composite flowers, common both here and in England, is known in my area as German weed, elsewhere as French weed. I have never heard a common English name for it. Botanically it is *Galinsoga*, named for an eighteenth-century Spanish botanist, Mariano Martinez de Galinsoga. After the worst bombing of London in World War II this weed sprang up among the ruins—hardy, green, and, Londoners said, gallant. At first unknown to most people, it was identified by botanists, and soon the name *Galinsoga* had been exalted, not corrupted, into gallant soldier.

I can think offhand of only two plants that honor people by their given names. Timothy, the hay grass, was named for farmer Timothy Hanson, who introduced it from New England to the southern states early in the eighteenth century. The other is Joe-Pye weed. The botanical name, *Eupatorium*, honors Mithridates Eupator, king of the ancient country of Pontus until 63 B.C. who was said to use it as a medicine. So did the legendary Joe Pye, an Indian herb doctor in New England. Only a few years ago, the first authentic records of Joe Pye were found in an old tavern account book from Stockbridge, Massachusetts, where Joe bought rum on tick.

And finally, since one cannot go on and on with such lore, there are the out-and-out folk-poetry names, some quaint, some droll, some wry, some beautiful. Like blue-eyed grass and stargrass. Or pennyroyal, bluecurls, bugleweed, skullcap, and self-heal, all members of the big mint family. There are bilberry, whortleberry, sparkleberry, farkleberry, and deerberry—all kinds of blueberries. And the countless viburnums—arrowwood, wayfaring tree, hobblebush or tangle-legs, wild raisin, stagbush, nannyberry, dockmackie, pimbina or mooseberry. There are dogbane and wolfsbane and baneberry, and all the other banes. Plus beggar's-lice or hound's-tongue, squaw-weed, butter-weed, stinking Willie, virgin's bower, monkshood, toadflax, butter-and-eggs, lamb's-quarters, pokeweed, mugwort, smartweed, tearthumb, cudweed, everlasting, goldenrod, and eyebright. As well as bluets, also known as Quaker-ladies or innocence. And, of course, ragweed, *Ambrosia* in the Latin and perhaps the most flagrantly misnamed plant in the whole of botany.

We need the scientific names, of course, and we couldn't do without them. But we live with and speak the common names every day of our lives.

The Running Country

PHOTOGRAPHS BY PATRICIA CAULFIELD

TEXT BY JOHN MADSON

LIEUTENANT COLONEL Stephen Kearny's orders were clear: proceed northwest into the Iowa wilderness with a unit of cavalry and survey the Des Moines River for possible fort sites. They left Keokuk in June 1835 with a couple of transport wagons and a small herd of beef cattle, riding along the crest of a sun-drenched prairie ridge between the Des Moines and Skunk rivers, bound upstream into unknown country.

It was a beautiful time, as only an early prairie summer can be beautiful. The ridges glowed with flowers, and when the wind parted the new grass ahead, Kearny saw wild strawberries "that made the whole track red for miles" and stained the horses' hooves and fetlocks.

Slowed by the wagons and cattle, the party was averaging fifteen miles each day. As it turned out, that was about the same rate that the strawberries were ripening; the soldiers and the berry ripening were traveling north together. As if that weren't lucky enough, one of the cows freshened and began giving milk, and the troop dined on fresh strawberries and cream all the way to the headwaters of the Des Moines.

First-class foraging, and it sure beats beans. But then, this strange new country beat almost anything.

It was virgin tallgrass prairie, and Kearny and his men rode stirrup-deep through young bluestem grass and flowers, the hooves muffled in a loamy wealth that had been accruing annual interest for twenty thousand years. Later that summer, when they skirted ridges and crossed flats, their horses would vanish in a sea of Indian grass and big bluestem so tall that it could be tied in knots across the pommel of a cavalry saddle. It was a land belonging to grass, flowers, and sun, a new sort of land that was open to the sky, and trees and shadows shrank from it. For a long time, so did people.

The first signs of prairie began back in Ohio as little natural clearings in the great eastern

156

hardwood forest. They were strange clearings filled with strange grasses, and they saved a lot of axework. These eastern outriders of the prairie were quickly filled with fields and people, and vanished before men had a chance to really know them.

Scattered openings continued across northern Indiana, becoming larger, although the land was mostly forest. And then suddenly, twenty miles west of the Wabash River, the world opened up. A man would walk up out of the forested floodplain, step through a screen of sumac and wild plum, and stand blinking in a land that blazed with light and space. He was at the eastern edge of the Grand Prairie of Illinois; from there, north to Lake Michigan and west to the Mississippi, the prairies opened and broadened, sometimes spanning fifty miles without a tree or any other object to break the fabric of the grassland.

From the Wabash, this tallgrass prairie ran to the Missouri River and beyond, covering the western parts of Missouri and Minnesota and almost all of Iowa, extending into the eastern Dakotas, deep into Nebraska, and down into Kansas, Oklahoma, and Texas. It was called *tall-grass prairie* because it was a region dominated by huge grasses—Indian grass, the cordgrasses, and big bluestem, which might grow twelve feet high. It was a special region, labeled clearly and precisely with special plants. At its western boundary, out around the 100th meridian, the tallgrass prairie merged with shorter mixed grasses and midgrasses, which merged in turn with the short grasses of the Great Plains.

True prairie was not a matter of location, but of composition. The lie of the land had nothing to do with whether it was prairie or not; if it was tallgrass prairie it included the tallgrass communities. Some prairie was flat, much of it was rolling, and some was broken and rocky. But it needed tallgrasses if it was to qualify as true prairie—the most easterly of the great American grassland societies that sprawled between the Rockies and the eastern forests.

It was here that the forested East ended, and the West really began. It stunned the pioneers coming from their Ohio and Kentucky forests, and one old journal effused: "The verdure and flowers are beautiful, and the absences of shade and the consequent profusion of light, produces a gaiety which animates every beholder."

Open as it was, it was not treeless. My home country in Iowa was a series of named prairies, such as the Ross Prairie or the Posegate Prairie, or whatever, and these stretches of shaggy grass were more or less fenced with groves and timbered valleys. Such prairies were said to resemble lakes, with boundary timber as shorelines indented with deep vistas like bays and inlets, throwing out long points that were capes and headlands.

Years ago, when Americans were less homogenized and outlandish accents still drew attention, a man in Maine asked me where I hailed from. I told him I was from Iowa. He shuddered, and calc'lated that his forests and hills must seem very beautiful to me. I replied that it was sure different from anything back in Story County, all right, and that seemed to please him.

158

The truth of it was, my home country had about all the hills and trees that we needed.

Our prairie country had a marked pitch and roll to it, like an ocean quieting after a bad storm. There is prairie country that's about as flat as land can be, but most prairie has a fine roll and break, with the land billowing off to the skyline and some timber down in the folds.

Many of the original trees were the same as back East: elms, hard maples, silver maples, shagbark hickories. But in Iowa, the beech tree's place was taken by basswood; I never saw a beech until I was thirty years old.

The prairie forests varied, depending on where you found them. Along smaller streams were wild plum, box elder, wild cherry, soft maple, elm, and wild grape. If the floodplain was flat and wide, there were walnuts, hackberries, and great cottonwoods. These floodplain forests were densely undergrown, but this wasn't true of forest higher up on the ridges. There the trees were encroaching on prairie domain, and had to recognize the sovereignty of King Grass. These upland forests, what there were of them, were open. The groves of oaks and maples, huge-boled

159

and ancient, looked like royal parks. The ground beneath them was free of undergrowth and carpeted with shade-tolerant grasses and flowers. Those old upland groves are gone today, mostly, but now and then you'll see a sentinel oak at the edge of a high pasture. After a hard rain, the closely grazed ground on such a ridge may glitter with flint flakes—the tribes knew good summer camps when they saw them.

Primeval prairie woods were found on sand and clay ridges, rocky outcrops, or the floodplains of streams. All the richest parts of the original prairie country were in grass; the forest existed at the sufferance of grass, and only on places that grass did not choose to occupy.

I've never thought of such country as monotonous, although my Maine friend would have felt so, I'm sure. But I've had four generations to get the forest out of my blood. By 1800, some of my people had begun to peer warily out of the Ohio forests; by 1825 they had gotten as far west as the eastern edge of the Grand Prairie. Bit by bit, they crept out of the trees and weaned themselves away from the Wooden Country. In 1853 my great-granddad grasped the

*Down in a wet prairie swale, with its feet in water, is a wind-whipped stand
of slough grass, also called "rip-gut" because of its tough, saw-edged blades.
Its sod was unexcelled for the "soddie" homes of early settlers. When cut early,
it also made good hay. But the glossy blades didn't stack well, and a full load
of slough grass hay had a maddening habit of slipping off a wagon.*

nettle (and a new bride) and left the old states, heading west into the prairie frontier of central Iowa.

The strain has bred true; I like to return to trees, and sit and walk and hunt under them, but I could never live under them if they kept out all the sky.

Those royal groves must have been something, though.

Why all that grass?

Why, suddenly, twenty miles west of the Wabash, did the land begin running out of trees?

Some early settlers thought the land was just too poor to grow trees—but it didn't take them long to find that wasn't true. They finally decided that prairie was caused and maintained by the fall and spring infernos that swept through the grasslands, leaving leagues of blackened ash and carbonizing any tree seedling that had the temerity to invade prairie.

There was a lot to this. Certain islands in prairie lakes had fine groves of ash trees where fire could never reach. Yet the fire theory had gaps in it, and some early ecologists began to suspect that fire was an effect of prairie rather than a cause.

Mostly, it was a matter of rainfall. Eastern forests are in humid climate while grasslands are in drier country. The plains and prairies lie in the rain shadow of the Rocky Mountains—a lofty barricade to the moisture-laden winds from the Pacific. Prevailing winds carry enough rain to grow grass, but not unlimited forests. Then, too, prairie maintains itself well. Even at the eastern edge of the grassland, where rainfall was sufficient to support either tallgrasses or trees, the dividing line was abrupt. Tallgrass prairie is a closed community that rarely admits aliens; tree seedlings can seldom live in prairie sod with its intense competition, crowding, and ground-level shading.

Years ago, a pioneer ecologist named Bohumil Shimek began to suspect that the grasslands weren't just caused by low rainfall and maintained by fire, but resulted from evaporation caused by incessant exposure to wind, low relative humidity, and frequent high temperatures.

One of the greatest of these, Professor Shimek felt, was wind. Wind breaks twigs and leaves, and drives dust and sand against delicate tissues, abrading and tearing them. The effect of constant wind on tree leaves is also physiological. Shaking a plant increases its rate of transpiration—in a climate where a tree can't afford to transpire too much water vapor. Of course, this is checked by the closing of the leaves' stomata. But that also checks the processes of respiration and assimilation. If physical shaking continues long and violently, the plant can be weakened and even killed.

It follows, then, that trees most exposed in their spring and summer leaf seasons to hot, constant winds would be in the greatest danger.

This is apparent in Iowa, where spring and summer winds often blow from the southwest. Western Iowa streams that drain into the Missouri River flow southwest, directly into these winds. Eastern Iowa streams flowing into the Mississippi run southeast—at right angles to the prevailing winds and with maximum protection. Iowa streams flowing southwest may have scanty, brushy timber if they have trees at all, and some flow for miles through open prairie. The Maple, Little Sioux, Boyer, and Nishnabotna are examples. In eastern Iowa, the southeast-flowing streams—the Cedar, Wapsipinicon, Des Moines, Iowa, and Skunk—have dense floodplain forests.

In western Illinois we can see the effects of wind and sun on the hills above the Mississippi. Certain forested headlands and bluffs above the river are capped with little tallgrass prairies. You won't find these on east-facing hillsides, looking away from the river. The hill prairies face west, at right angles to the full blaze of an August afternoon, looking out over a broad floodplain where the wind has a long fetch. Of all our Illinois landscapes, none are more exposed to intense sun and wind than these west-facing river bluffs and hills. They are ecological niches of native prairie.

The steep ground that creates these prairies may also protect them. Some of the limestone river bluffs near my home are nearly two hundred feet high, towering over the Great River Road and the Mississippi in white, buttressed walls. The edges of these cliffs are dangerous pastures, as some farmers have found. Cows, unlike cats, can't land on their feet nor spare eight lives. Just behind these river bluffs may be rough, wooded valleys that discourage any farming. Guarded from front and behind, the little prairies along the brinks of these river cliffs have survived.

Our family often goes up there for Sunday lunches, climbing gargantuan limestone stairs to the clifftop. With forest at our backs and the broad Mississippi out in front, we bask in original bluestem high above the beaten highway of a newer, noisier world.

The old-time prairie was a grandmother's quilt of color and form that shifted constantly as the wind breathed life into the grasses. Willa Cather remembered Nebraska when "there was so much motion to it; the whole country seemed, somehow, to be running." One of Harvey Dunn's finest paintings of prairie life was of a Dakota girl pumping water, her skirts blowing, an embodiment of the old prairie adage: "There's just nothin' prettier than a girl pumping water in the wind!"

The play of wind on tallgrasses, with the land running beneath a towering sky, is something we may not know again, for we will not see such vistas of grass again.

The tallest of the prairie grasses—big bluestem and cordgrass or slough grass—never reached their highest growth in the richest soil, but in lower, marshier land. There, where substrata of clay lay near the surface, the big bluestem grew to twelve feet. Cows could be lost in it, and might be found by a mounted man only when he stood in the saddle or rode up a nearby ridge and watched for the cows moving through the deep grass. The size of prairie grass was proportional to moisture. The dark green slough grass, called "rip-gut" because of its saw-edged leaves, grew

in dense stands on low flats. Pioneers avoided low prairie swales that were marked by this "black grass." A traveler had to pick his way carefully over the ridges in spring, for the low places were impassable. When incumbent frontier politicians made soaring reelection promises, they often swore that they had "waded sloughs" in the interests of their constituents, for no work was harder than that.

Another lowland grass was Indian grass—tall, coarse, and up to eight feet high, usually found in more southerly regions of tallgrass prairie. All of these are fine livestock feeds, and make excellent hay if cut before their stems are too tough and fibrous.

The higher and better drained the prairie became, the finer and shorter the grasses. True prairie uplands are dominated by little bluestem, with rich stands of Junegrass, side-oats grama, needlegrass, and prairie dropseed.

The tallgrasses need moisture and plenty of it, and were the dominant grasses in the prairies east of the Mississippi. Farther west, as rainfall diminished, tallgrasses retreated to lower parts

The little wetlands of the tallgrass prairie were as rich as the land around them, and they were crammed with such aquatic plant life as the long-leaved Potamogeton pondweeds and tiny, free-floating duckweed.

of the prairie, and the well-drained uplands were covered with shorter midgrasses. Still farther west, with even less rainfall, those midgrasses were replaced by the shortgrasses of the true plains: western wheatgrass, buffalo grass, and blue grama.

Colors and textures of tallgrass prairies varied with the season and elevation. In early spring, the bleak prairie hillsides might be brownish gray from the weathered ash of the fall fires. Then, often well before Easter, the prairie pasqueflowers would appear. Of all early wildflowers these are the bravest, not blooming in sheltered woods, but out in the big open on glacial moraines where the wind cuts to the bone in late March.

Then, one bright morning, the south-facing slopes would look as if patches of spring sky had fallen on them, and you knew that the bird's foot violets were in bloom. There was white, woolly, prairie cat's-foot coming on, and the first green blush of new grass on the slopes.

The hilltops were splashed with early spring flowers: false dandelion, cream-colored paint-brush, and mats of groundplum vetch. The new grass would be spangled with tiny purple, blue, and white grass flowers, and perhaps yellow upland buttercups and yellow lousewort.

When the bluestem grasses began to appear in mid-April it was a signal for the spring flowers to hurry, for they were small plants that were easily overpowered by the growing grasses.

Prairie and meadow violets appeared, with vetch and false indigo. Along the streams and low places were marsh marigold, yellow stargrass, and purple heart-leaved violets. The prairie pinks came into bloom and enameled a landscape of young grass with pink, white, and purple. With them came the puccoons, splashes of rich orange in the greens and pinks.

By early June, most of the spring flowers were gone. The flowers were taller now. Daisies began to appear, and larkspur and purple coneflower. There was a foot-high prairie lily with a red bloom, and with these lilies came clouds of prairie roses. About the only thing that "them politicians down at Des Moines" ever did that pleased our Grandma Tut was to make the prairie rose Iowa's state flower.

By summer there were myriads of blossoms, all holding their own with the lofty grasses. Wild indigo, with its heavy, creamy blooms, stood tall. White larkspur stood above many grasses; so did oxeye daisy, many sunflowers, goldenrod, and compass plant—the set of its oaklike leaves marking the prairie meridian. Leadplant, with its silver-gray leaves and purplish flowers, was everywhere on the upland prairie. From August on through autumn, wild asters bloomed white, lavender, and purple. Deep in the ripening grasses, almost hidden, were the fringed gentians and bottle gentians.

This was the season when the prairie flamed with blazing-star or gayfeather, a tall purple spike of blooms whose root bulbs were fed to Indian ponies to increase speed and endurance.

Some of the finest floral displays were on low ground, hidden around the prairie marshes. Pioneers recall four types of wild orchids there. The smallest, and possibly the rarest, was yellow. There was a white orchid with purple mottling, and a larger yellow orchid. The largest and finest of the wild orchids in Iowa grew two feet high far back in the marshes; on each stalk there

Old Sitting Bull himself, in the dry, flaming autumn of 1885, warned some Dakota school-children that they could never run away from a prairie fire. "Go to bare ground," he counseled them, "or onto sand, gravel, or plowing. Or set a backfire. Go to a place with no grass. But do not run."

Entire towns were destroyed by some of these prairie fires; in Leola, South Dakota, all but twelve of the town's hundred buildings were burned in 1889 by a prairie fire that traveled forty miles in four hours.

Prairie fires were feared by almost every pioneer, and even the small boys who usually found them exciting had to admit there were drawbacks. Herbert Quick, the Iowa writer, told of the sharp grass stubs on burned-over prairie that pricked a schoolboy's bare feet and caused festering sores. Of course, this could have been solved by simply wearing shoes to school, but that's a stupid solution when it's spring and you're ten years old and the prairies are greening up. Those prairie boys found a friend in need in the pocket gopher. The big gophers threw up mounds of soft, fine, cool earth, often in long lines. The ultimate luxury was to walk all the way to a country school without stepping off a soft gopher mound. It never worked that way, of course. The gopher mounds usually wandered off in the wrong direction and a boy ended up walking farther to school through grass stubs than if he'd taken a direct route in the first place.

That problem, like all others caused by prairie fires, vanished with the bluestem. There came a time when the autumn horizons no longer glowed red at night, for the prairie was gone.

We spent our tallgrass prairie with a prodigal hand, and it probably had to be that way, for these are the richest farm soils in the world. There were certain wilderness things that were fated to be spent almost to the vanishing point: bison in shortgrass plains, lobos and grizzlies in settled cattle country—and the vistas of true prairie.

But spending is one thing; bankruptcy is another.

To squander the last stands of true prairie would wipe out a valuable index to original quality. It is important that our agronomists, botanists, zoologists, and soil physicists have reference points to the original plants and soils of our most valuable ecosystem. We may someday have to rebuild those soils, or try to. Native prairie is a base line from which creative research can depart, and return for reference. We'd never dream of melting down the platinum meter in Paris and converting it to jewelry; it is the master rule, an original measurement upon which so much engineering and science are based. And so, in an even greater sense, is native prairie.

Just as important is the maintenance of certain tangible links with the old time. To destroy the last of the native prairie would be as criminally stupid as burning history books, for prairie is a chronicle of human courage, endurance, and victory, as well as a finished natural system.

Deciding sometime tomorrow afternoon that maybe we should have a few more prairies around, and then abandoning a few cornfields to that end, just won't get it. Original tallgrass prairie is the end point of twenty-five million years of evolution; it cannot be restored overnight, if indeed it can be restored at all. If the job of prairie restoration were left to Nature alone, and if there were adequate sources of seed, Nature might be able to convert cornfields to "native" prairie in two hundred to three hundred years. Imitation prairies have been built, but with perhaps thirty plant species instead of the original two hundred or more, each occupying a special niche in a special way. Prairie in its full form cannot be reconstructed by man—Eric

Hoffer notwithstanding.

However, tallgrass prairie is the most difficult of all native America to conserve. This is because it is the world's most valuable farm soil—and it must be conserved in quantity if it is to mean what it should. There are still many "splinter prairies" in Midwestern states, some of several hundred acres. A man could stand in a grove of virgin white pine of the same size and feel that he was in primeval forest. Not so with prairie. To the average man, a scrap of native prairie is just a shaggy weedpatch between cornfields. Prairie must have sweep and perspective to look like prairie. It is more than just native grasses and forbs; it is native sky, and native horizons that stretch the eye and the mind. To be prairie, really good prairie, it must embrace the horizons. That is the ideal, and the only places where you will still find it are in parts of Nebraska's Sand Hills and in the Flint Hills of eastern Kansas.

The Flint Hills prairie has survived because beds of cherty limestone lie so close to the surface that the land can't be plowed. It is heavily grazed, and has been for a hundred years, but it is still prairie, rolling in long waves from the Nebraska line down into the Osage Hills of Oklahoma. It's country worth seeing.

One morning in early April, Kansas biologists Bob McWhorter and Bob Henderson hauled me out of bed at 3:00 A.M. for a prairie chicken count in the northern Flint Hills.

Before dawn we were on the ancient dancing grounds of the prairie chicken, with a traditional stage setting. The southern horizon was flame-torn with spring prairie fires that reddened the sky and gave the strange impression of sunset at dawn. Somewhere behind us several coyotes were swapping hunting yarns and settling down for the day, and even before it was light enough to see, we could hear the haunting, hollow booming of prairie chickens from several directions, some nearby and others dim with distance. With first light we could see them on the prairie ridge before us, over thirty of them in the closest flock, the sun glowing orangely on the inflated air sacs of the dancing, posturing males. As we watched, a phalanx of upland plovers swept past. Just over the ridgetop to the north, not twenty yards above the fire-blackened prairie that was beginning to blush green with new bluestem, a column of Canada geese moved out for breakfast. Another prairie day had begun, much the same as spring days have begun on Kansas prairie since the Miocene.

There is a proposal to reserve a great block of this Kansas prairie and create a Prairie National Park. It has all the elements—a splendid roll to the land, most of the original plants and animals, and enough physical dimension to look the way prairie ought to look.

The National Park Service is intensely interested in such a park, but wants a mandate from Kansans themselves, and Kansans are split on the issue. Cattlemen are generally against it, and you can't blame them. Outdoorsmen and naturalists are in favor of it, and tourism promoters are wild over the idea because it will give them something to promote for a change. Kansans are touchy about the fact that they haven't any spectacular attractions to shortstop some of that tourist money on its way to Estes Park, Colorado, and they have been reduced to putting up signs at state entry points reading: "Welcome to Kansas—Home of Beautiful Women." There's a lot to that. But most Kansans feel that their beautiful ladies would be enhanced with the backdrop of a Prairie National Park.

We need the big park, but we need the remnants, too. There are still many tallgrass prairies, shrunken and fragmented, surviving in scraps along old railroad rights-of-way, neglected fence

173

corners and farm lanes, and hill prairies in forest openings. Knowing their value, certain colleges and state conservation departments are doing all they can to acquire and preserve relic prairies. Lucky the school with such land!

In my home town of Ames, Iowa, high school biology teacher Dick Trump has a teacher's dream behind his classroom. He can lead his class out the back door, across a service drive, and through a tall gate with the sign: "Ames Senior High School Study Prairie." There, on the crest and slope of a hill, is a seven-acre patch of native Iowa prairie. It has about everything that native prairie should have—plus a very wise teacher who knows an academic windfall when he sees one. For a while, there was danger that the little prairie would be used for a new school ware-house, and local conservationists have breathed easier since the school board agreed to lease the school prairie to the Nature Conservancy for forty-nine years.

Of the relic prairies that I know, none is as poignant as the tiny scrap that I found years ago in the center of an intensely farmed Iowa section.

It was a small, lost graveyard, all that remained of the little settlement of Bloomington, wiped out by diphtheria over a hundred years ago. About a dozen weathered stone markers leaned and lay in a patch of original bluestem. Among the graves were those of a young mother and her children, and when I found the place in late summer their graves were set about with a few tall, magenta torches of blazing-star, stateliest of all prairie flowers. It was part of an original time and place, and it held fitting memorials. There were the flowers of gayfeather to lift the spirits of lonely, beauty-starved women. There was bluestem for the men, for their haycutting and prairie chicken hunting. For the children there was compass plant, with its wonderful chewing gum, and wild strawberries hidden in the grass.

That patch of tallgrass prairie was a more enduring memorial than the stones that stood there, and infinitely more appropriate. Today, our memorials reflect our values, and we will probably be interred in manicured "memory gardens," our graves decked with plastic blossoms that are imitations of imitations. That, too, may be appropriate.

My feeling for tallgrass prairie is like that of a modern man who has fallen in love with the face in a faded tintype. Only the frame is still real; the rest is illusion and dream. So it is with the original prairie. The beautiful face of it had faded before I was born, before I had a chance to touch and feel it, and all that I have known of the prairie is the setting and the mood—a broad sky of pure and intense light, with a sort of loftiness to the days, and the young prairie-born winds running past me from open horizons.

A strong place puts a mark on all that lives there, and the mark may outlast the place itself. Prairie people are like their western meadowlarks, seeming to be the same as their eastern rela-tives, but with a different song. It was the prairie that changed them all. It gave them a new song, and new reason for singing.

Compass plant or rosinweed, beloved by pioneer children for its wonderful chewing gum,
marks the meridians of the prairie with the set of its huge, oaklike leaves.
The stiff stalks may stand ten feet and are favorite perches for bobolinks, dickcissels,
and meadowlarks. Early settlers, trying to travel on higher ground to avoid sloughs,
sometimes marked wagon routes by tying bits of cloth to the tops of compass plant stalks.

Chaparral

PHOTOGRAPHS BY DENNIS BROKAW

TEXT BY SYLVIA FOSTER

To MANY people it is just that "scrubby stuff," the brush that grows, dense as a mat, on the hills of southern California. But to others, who have come to know it intimately, to appreciate its special beauties and unique qualities, the chaparral is more. Much more. Specifically it is a highly specialized type of brush, but to say it is that, and nothing else, is like saying the redwoods are merely trees, and nothing else. A definition of the chaparral is incomplete without recognizing that it is also the land itself, and the changing moods of that land, its colors, sounds, and fragrant scents. Chaparral is the annual spring show of ceanothus, the "wild lilac," dressing the countryside in a profusion of delicate blue (or white) blossoms. It is the pungent scent of sage, wild as the wildness itself, released from the aromatic leaves by a bruising winter rain. It is a pack of coyotes, in a blue-veiled twilight, giving frenzied voice to their hunger as they course after a terrified jackrabbit. It is the deep, all-pervading silence of the night, broken by the disembodied cry of the poor-will echoing down remote, dark ridges. A disquieting sound, it seems forever repeated, forever—*poor-will, poor-will, poor-will-ip.*

It is solitude and space unbroken, where the treeless hills roll like visible drumbeats from the mountains to the sea.

Chaparral—a resonant and romantic word, it glides over the tongue like a murmur from California's colorful past, vividly bringing to mind a time when this singular brushland, unshorn by the bulldozer, graced the rugged mountain slopes and mesas of southern California from the Pacific Ocean to the lower reaches of the Sierra Nevada, from Shasta County south into Mexico. In those early days herds of cattle roamed the grassy hills and brushlands unhindered by fence or freeway. Hardy *vaqueros* tending them wore leather shields to protect their legs from the ravages of the brush, through which they often had to force their passage. A particular nuisance was a thorny little oak known to them as *chaparro*—the scrub oak, *Quercus dumosa.* It was a token of their begrudging respect for this little oak's nearly impenetrable thickets that their leather leggings ultimately were referred to as *chaps,* and the brushland itself —chaparral.

176

On a great boulder, the spiny leaves of live oak, damp with rain, greens fading to gold and russet and mould.

Between yesterday and today the vast domain of the chaparral has been drastically reduced. Land has been needed for industry, for agriculture, and for homes to accommodate southern California's ever-growing population. Nevertheless, virgin stands still maintain a deep-rooted, tenacious hold on several million acres, making the chaparral the most extensive forest-type cover in California. And, though the comparison may seem absurd, the chaparral is indeed a forest, a forest in the sense that it takes the place of, and serves similar functions to, the usual trees that one envisions when the word forest is mentioned. It can be regarded then as a pygmy forest, whose thick cover protects the watershed, thus preventing rainwater from running pell-mell off the steep southern California hills, taking with it precious topsoil, and otherwise causing widespread erosion and flooding.

As a forest cover it also offers food and refuge to its surprisingly abundant and highly varied wildlife population. Mule deer, coyotes, fox, bobcat, rabbits, and innumerable rodents from tiny shrews to kangaroo rats, wood rats, and ground squirrels are among those mammals who find food and shelter in its dense thickets. Scores of birds, from residents such as the pale-eyed

wrentit—a unique species peculiar to the chaparral and lowland thickets of the West Coast ranges—to migrants including colorful Audubon's warblers and lazuli buntings, find the chaparral to their liking, particularly in the winter and spring when ample food and moisture are available.

As a source of food and fiber for man, the chaparral forest is of little economic importance today. Nor was it particularly important to the early settlers, except perhaps as fuel for their fires. But to the Indians of southern California—and later to Spanish-Americans who adapted Indian uses to their own needs—many of the chaparral shrubs were indispensable. Notable among these was the chaparral's namesake, the indomitable little scrub oak, whose branches yielded hundreds of pounds of acorns to be ground in their *morteros*. Berries of the manzanita—Spanish for "little apple," so named for the shrub's tiny, apple-shaped fruits—were harvested, as were the fruits of the holly-leaved cherry. Tough but pliable branchlets from the squawbush, first cousin to the disreputable poison oak, were woven into hardy baskets. And yerba santa, the "sacred plant," highly prized for its medicinal properties, was used as a remedy for sore throats and lung ailments.

These shrubs which the Indians valued so highly, and utilized so judiciously, are several of a mere twelve genera considered as "true" chaparral. That is, they are the shrubs recognized by many botanists as being dominant within the chaparral's limits. Though few in number, all have several characteristics in common which readily separate them from any other plant community. The most noticeable is their size. They are typically dwarfed, seldom attaining a height over twenty feet, and often look like small, rounded trees. Their leaves, adapted to conserve moisture, are small, thick, and leathery and of a dull-green color. These various shrubs grow thickly together, with limbs and branches so greatly intertwined that the only mode of travel, in many instances, would be to follow a game trail on hands and knees, and even then with difficulty.

Many other plants, both shrubby and herbaceous, thrive in the chaparral, sheltered by its hardy, deep-rooted growth. These include annual wildflowers, and such beautiful native plants as the vividly colored penstemons, the red bush monkey-flowers, and the spectacular yuccas, "candles of our Lord," all of which provide color and accent to the outward monotony of the chaparral. Together with these associated plants, the chaparral forms a highly specialized community found only in the Southwest, although similar-appearing plant formations can be found in Australia, southern Europe, along the west coast of Chile, and in southernmost Africa.

With its characteristic structure and growth habits, the chaparral is perfectly adapted to the semiarid conditions under which it must exist. Southern California, open to the sea for two hundred miles, has a climate which is tempered by moist, marine air. Days are usually mild and sunny. Annual rainfall, predominantly in the winter months, is moderate. Hence seasonal changes, in the conventional sense, are barely distinguishable, and two seasons, rather than four, are more generally recognized. There are many variations, of course. An inland valley will have higher and lower temperatures, and perhaps more rain, than will a coastal plain. But typically, the first season, from April to November, is long, arid, often hot, with frequent desiccating winds, while the second, from November to April, is short, moist, and cool. Hence the growing

season, "spring," comes to the chaparral during late winter, when ample moisture is available, and "winter dormancy" during the summer, when there is a dearth of it. A unique cycle to which the chaparral is singularly adapted, it is unlike most "normal" forest types which, because of their more typical growth cycles and greater moisture demands, could not possibly survive under these same conditions.

This cycle, unique as it is, puts the chaparral at an esthetic disadvantage, a disadvantage quickly noted by newcomers to southern California, fresh from moist, green eastern woodlands or northern forests. For during the summer, when other forest covers are at the peak of their growing season magnificence, the chaparral is easing into unattractive dormancy. Generally, by midsummer, most of the shrubs and herbaceous plants of the chaparral have flowered and fruited, and their seed wind-sown or distributed by birds or other small animals, who harvest them for food. Annual wildflowers and grasses have withered, and disappeared utterly from the landscape, leaving only brown stubble to mark their passing.

The hills become suffused with the softly somber tones of an old tapestry, dull red-browns, old golds, faded greens, and grays. This somberness, typical of chaparral summers, is partially a result of seasonal dieback, standing deadwood, and the fact that some shrubs and herbaceous plants—ceanothus, mountain mahogany, black sage—drop part of their leaves, thereby acquiring a "lifeless" cast as lack of moisture naturally brings about dormancy. One shrub which belies this apparent lifelessness is the toyon, the California holly or Christmasberry. Its sprays of white flowers are shortly followed by clusters of berries which, when red-ripe during the winter, will attract hordes of visiting robins, western bluebirds, and cedar waxwings to its laden branches. Now, however, just the *chir-r-r-r* of cicadas, the scurrying of lizards in the brush, and the calls of resident birds such as the melodic California thrasher break the hushed and shimmering monotony of the chaparral days.

As the sun shines relentlessly for month after rainless month, the chaparral becomes increasingly dry and desiccated until the hills become flammable as a match. Fire is not just a threat. It is a foreseeable disaster. Frequently started carelessly by man, a fire can be fanned by a fast-paced Santa Ana wind into devastating rage in minutes. The annual loss of thousands of acres of valuable watershed, and millions of dollars worth of private and public property, to fires in southern California is inestimably tragic. More tragic is the loss of wildlife, and occasionally even human life, that occurs in these dreadful holocausts.

But, dreadful as they are, fires are nothing new to southern California; they have been ravaging it for untold years. Dr. Willis Jepson, eminent California botanist, has called the chaparral a "fire-type" formation. That is, while it is subject to recurrent burns, it recuperates vigorously from such devastation. Some shrubs, such as chamise, or greasewood, respond to burning much as a garden ornamental responds to severe pruning and, along with scrub oak and the nearly indestructible laurel sumac, sends up new shoots from surviving crown and root stock in a surprisingly short time. Dormant seeds, lying buried under a protective layer of earth, are triggered into sprouting by the heat. Still others, brought in by wind or by birds, take root beneath the ashes.

Come spring, and a good rain, a host of wildflowers—phacelia, lupines, snapdragons, poppies, gold fields—seemingly burst into bloom overnight, their flagrant color an astonishing contrast to the blackened earth and charred skeletons of the shrubs. This profusion of flower and leaf,

The shrubs of the chaparral "grow thickly together, with limbs and branches
so greatly intertwined that the only mode of travel, in many instances,
would be to follow a game trail on hands and knees, and even then
with difficulty." Like this laurel sumac, "they are typically dwarfed, seldom
attaining a height over twenty feet, and often look like small, rounded trees."

this seemingly spontaneous growth, soon covers the hills with vibrant color, and lush, low greenness. But the flowers are only temporary, merely the finery of spring, and the root and crown growth is not the entire shrub. For the shrubs to grow entire, for the chaparral to again reach a full climax stage, twenty to twenty-five years must pass.

The first rain, usually in late November, ends the extremely hazardous fire conditions, and simultaneously initiates the second, shorter "wet" season. Heavy rains, often torrential, will continue periodically through late spring. With this welcome moisture the shrubs begin to respond to the compelling force of growth in their root systems, but will not bloom until late winter or early spring. When "spring" does come to the chaparral it seems to come unannounced. The inattentive might miss it entirely, for this, the chaparral's period of greatest beauty, is short and sweet, the annuals and some shrubs and herbaceous plants blooming and dying back often in a matter of weeks.

In a sometimes discontinuous, often overlapping order of succession, beginning with manzanita in January and ending with toyon in June and July, the chaparral arrays the hills of southern California with bloom. Masses of ceanothus, delicate blossoms, sky-blue to almost purple, or white, nearly enveloping the shrubs themselves, sprawl down the slopes. Mountain mahogany, whose feathery styles lend a silvery luster to its branches, shares honors on a warm embankment with laurel sumac, whose rich magenta primary leaves are dazzling forerunners to its later, purplish-red buds and loose, cone-shaped bunches of tiny ivory flowers. The buds of chamise, like minuscule beadlets, burst ultimately into feathery, spiraealike sprays and dapple the hills with patches of creamy blossoms.

Aromatic bush rue, with its citruslike flowers and red-golden berries, and black sage with its blue bee-tended flowers, pleasure the senses, as does the sugar bush with its sticky, sweet, flat-headed buds, which soon unfold, forming small, red-tinged white flowers. Lanky *Mimulus* (monkey-flower) spill their scarlet faces carelessly over a slope; yuccas, upthrust through the thickets, stand in stately, isolated beauty against the hills; yellow rockroses and wild buckwheat, like ecru lace, pattern the open areas on the fringes of the chaparral.

Everywhere wildflowers signal the season: lavender and white shooting stars and canchalaguas drift across the grassy slopes; blue dicks nod heads on tall, stringy stems; zygadenes spike the rocky outcroppings; suncups set a meadow ablaze with yellow; mariposa lilies unveil their delicate tints of lavender, pink, and white in secluded, brushy places.

The earth, supersaturated from the lavish, seasonal rains, drains its excess into the rain-torn arroyos. Here the waters collect and, particularly on the north sides, annual seeps appear. Some, in deeply shaded pockets, remain through most of the summer, havens for resident wildlife. In other sections of the chaparral are natural springs, which never dry out as ultimately do the seeps. About these damp places grow more riparian plants such as willows and cattails. Here flat, primitive, shade-loving liverworts and thick-napped mosses cushion great, granitic boulders already heavily encrusted with yellow, orange, and charcoal-hued lichens. Gold ferns find sufficient soil to root in rock crevices, while wood fern and California maidenhair thrive in the humus-rich understory. Brilliant fuschia-flowering gooseberry cascades over the embankments, tangling with necklaces of pale-yellow honeysuckle, and masses of spring-fresh poison ivy, and with wild cucumber vines, heavy with spiny, pendulous green fruits, to form a transient jungle as impenetrable as Guiana. Instinct now compels frogs and toads to seek these hidden pools to

New growth pushes up from within the shelter of a brown and withered rockrose,
backdropped by manzanita, rich with ripened berries, the "little apples"
of its Spanish name. Once, before the reign of the bulldozer, the chaparral
dominated the mesas and mountain slopes of southern California,
from the Pacific to the Sierra Nevada, from Shasta south into Mexico.

lay their eggs. And California slender salamanders, wormlike little amphibians, who also have sought out these seasonal water holes, can later be found curled beneath decaying stumps in the moist meadows.

This amazing rapidity of growth, this lush verdancy, so typical of the chaparral, is short-lived. Soon the cycle will repeat itself. The wild cucumber vine will wither and break, its remnants adding to the humus below. Mosses and liverworts will crumble into brown dust. Gooseberries will drop their leaves, revealing their thorny, naked superstructure. Pools which formerly harbored a fecund, if temporary, life of their own will shortly evaporate, some becoming caked and shattered clay. With lack of moisture and increase in temperature, the shrubs and herbaceous plants will again become dry and pallid, and ease into dormancy. The long wait of summer once more will have come to the chaparral. Still, inherent in the shrubs of the chaparral, as with all plant life, is the promise of growth and beauty to come. This promise will be fulfilled again with the first, bountiful, life-giving rain. So the cycle continues; only the seasons change.

This is the chaparral. Yet it is not all there is to know or to say. For though it is true that the chaparral has attained its most perfect development in southern California, it does range, in varying degrees, beyond the lower half of the state south and east into Mexico and Arizona, and north into northern California and Oregon. And, while what has been recorded here can be applied generally to all topographical areas in the southwest in which the chaparral is found, each area in which a particular chaparral shrub is dominant—together with such factors as elevation, annual rainfall, temperature, and soil type—will determine the characteristics of that area. There will be many, many variations. This has been but one.

More richly varied in plant and animal life than most people realize, that "scrubby stuff," the chaparral, stands alone. An integral part of California's heritage, it deserves closer attention and study, deeper appreciation—and greater protection, the same protection that any area of natural beauty which is faced with ultimate eradication deserves. It would be to southern California's detriment if, on some future day, anyone would need to inquire, "What was here? What was the land like before . . . ?"

The bone-hard skeletal branches of a manzanita are a stark reminder of the fire that years ago devastated this hillside. But the chaparral shrubs recuperate vigorously, and even thrive after fire.

The view from the mountain . . . of cloud-to-cloud
and cloud-to-ground lightning from a squall line
over the Idaho mountain wilderness.

The View from the Mountain

PHOTOGRAPHS BY JOHN DEEKS

TEXT BY ELIZABETH N. LAYNE

S HEEP HILL lookout in Idaho's Nez Percé National Forest is one of a dwindling number of fire towers still manned by the U.S. Forest Service. A decade ago, at their peak, there were five thousand towers in operation. But the use of aircraft for fire surveillance has rapidly reduced that number to fifteen hundred.

Inefficient though the old way may be, John Deeks finds being a fire lookout the ideal job. For it offers wilderness, constantly changing weather—and the opportunity to photograph both. Thus on June 15, 1971, John and his wife Karin were dropped by helicopter on the 8,411-foot-high summit of Sheep Hill in the Salmon River Breaks Primitive Area. A packer normally would have brought them with their supplies on horseback the thirty-five miles from the nearest ranger station. But although it was mid-June, the trail was blocked by twenty-foot snowdrifts.

The fire tower—actually a two-story cabin—would be home for John and Karin Deeks for the next four months. There was a wood stove for heat, a gas stove for cooking, and a spring a mile's hike away. But they seldom had to go for water; there was usually an abundance of snow to be melted. Far below them, where the fires would burn, it was already summer. But atop the mountain, the seasons would change from winter into spring, summer, autumn, and back into winter before the Deeks left in October.

From their viewpoint, the young couple could easily see the entire width of the state of Idaho, from the Bitterroot Mountains in Montana to the Seven Devils on the Oregon line. To the east, beyond the first peaks, the land lay green and rolling, with steep breaks leading down to the Salmon River. In June the river ran brown from spring melt, but by midsummer it flowed blue and clear. At night John Deeks could look up at the sky with his binoculars and see the Andromeda Galaxy "as though it were next door."

The year before, Deeks had been stationed at a lookout in southern California where the view, on a rare clear day, ranged from the Salton Sea into Mexico, and far out into the Pacific. But a

Continued on page 197

188

The view from the mountain . . . of unusually sharp patterns in an altocumulus formation that suggests waves washing ashore, of sky-filling altocumulus that vanished minutes later, leaving blue sky for a hundred miles.

The view from the mountain . . . of an enormous cumulonimbus cloud reaching from the ground to 50,000 feet and stretching the width of Idaho, of a cirrus cloud which could inspire an ancient sun worshiping artist.

The view from the mountain . . . of a rainstorm directly over the Salmon River gorge, of a midafternoon corona and a mother-of-pearl display from ice crystals in an altocumulus cloud, of the sun's crepuscular rays illuminating dust particles through cloud gaps.

The view from the mountain . . . of an incredible sunset created by altostratus and altocumulus clouds and the smoke from a forest fire.

thick gray-brown layer of smog almost constantly blotted out the world beyond his tower. Indeed, by midafternoon on an average day, the bulk of Mount San Jacinto that rose to 10,805 feet only a mile away would no longer be visible. Spotting fires through the smog was difficult. "But all of a sudden you'd see the black-brown and red smoke coming up into the muck."

A forest fire lookout naturally spends a lot of time watching the horizon and the skies. And it was natural for photographer John Deeks to focus his camera on intriguing cloud formations, on such phenomena as coronas, on dramatic sunsets and soft moonrises, on towering thunderheads and lightning fireworks.

A lot of storms let go in the Idaho mountains, and a lot of forest fires are started by lightning. And a tower perched alone atop an 8,400-foot pile of rocks naturally attracts a lot of attention in a thunderstorm. "The surge of electricity goes right up through your body, your hair stands straight up, and everytime you touch your clothes there's a great shower of blue sparks."

Toward the end of the Deeks' stay on Sheep Hill, ominous black squall lines started moving in from the southwest, bearing eighty-mile-per-hour winds, battering hail, and lightning crashing down "like all hell was loose." John and Karin left the lookout during one of these storms. The helicopter couldn't make it, but packer Merlin Wilson got through and took them by horseback down to the ranger station where the road to civilization begins.

The view from the mountain . . . of a clearing sky after a storm as fog rises from the Salmon, of a moonrise on a clear evening.

An old leader bighorn ram, named "Blunt Tips" to describe
his huge horns with their smoothly rounded ends, exhales
into a stiff, frosty November breeze. The first rays
of a morning sun cast the shadow of his horn onto the mist
of his own breath and etch his form into the dark evergreens.

Bighorn Profile

PHOTOGRAPHS AND TEXT BY JAMES K. MORGAN

MOUNTAIN SHEEP were born of the Pleistocene, an age that emphasized Nature's spectacular and bizarre. Masses of ice encrusted entire continents, receded, and spread again. Warm, dry climates followed major glaciations, and vast grasslands colonized the recent soils. The stark and life-forbidding glaciers gave way to an orgy of abundance that invited colonizing species to a genetic splurge and set the stage for excesses of aggression, horns, and tusks. A fairyland of enormous tusked mammoths, large-horned giraffes, magnificently antlered deer, and giant horned bison roamed the Pleistocene.

Mountain sheep first appear in the fossil record as ox-sized giants, arising in Eurasia some two and a half million years ago and reappearing, slightly larger than their present size, toward the end of the Ice Age. The fossil record of their evolution, ground away by glaciers, tells us little of how their present form was reached. But we do know that mountain sheep inherited extravagant gifts from the quixotic Pleistocene: the massive horns that equal the entire skeleton in weight and the violent head-butting rituals practiced by rams to establish their status in the peck order.

Some forty races of mountain sheep evolved as they spread around the northern hemisphere to most mountains of Europe, Asia, North Africa, and North America. These range in size from the giant Siberian argalis, which weigh up to four hundred and fifty pounds, to such dwarfs as the eighty-pound Cyprus urials. Molded into spectacularly successful glacier followers by the alternating poverty and plenty of glaciations, mountain sheep were part of the large mammal fauna that spread to the North American continent during the late Pleistocene. Their herding social system that passed social traditions from generation to generation, and their "fermentation vat" digestive system that efficiently utilized dry, abrasive forage, allowed them to exploit the bounty of climax grasslands. They survived the depredations of Stone-Age man and recurring glaciations to become one of the most successful mammals of the Ice Age and one of man's first domesticated animals.

Continued on page 211

A bighorn ewe has spotted a coyote in the soft early-morning light and is watching
intently as it hunts for mice. Her lamb, sensing tension, has sought the security
of its mother's side. Bighorns show little alarm, only wary interest, at the presence
of coyotes. When the intruder has left, the month-old lamb bounds down the hill,
leaping over a rock, to rejoin its playmates in an exuberant game of follow-the-leader
or king-of-the-mountain. If they discover a lingering snowbank in a sheltered gorge,
they will spend hours dashing, sliding, pirouetting, sometimes joined in their frolic
by the herd's yearlings. Raising bighorn lambs is a community effort. One or more
ewes supervise all the lambs, although a mother will nurse only her own offspring.

Young bighorn rams forsake well-worn clifftop trails for the challenge of routes down or across a precipitous rock face. This ram is frozen in mid-leap on a path through jagged rocks that—though it seems nonexistent—is really well defined. What look like random notches or protrusions would become, if connected by a line, a zigzag sheep trail with direction and purpose. If a point of rock is too small for a ram to stand on, it just bounces to the next stop. And a bighorn will also plunge down a rock chute, hurtling from niche to niche in a controlled fall with the ease that comes from superb traction and balance.

Following pages: A ram's horns serve both as status symbols and as weapons of dominance. But only rams of nearly equal horn size must fight to determine status in the herd's pecking order. This three-way encounter involves two large-horned rams and the small ram at the left. Victorious in this high-country skirmish, the ram on the right is shoving and bullying his now subordinate opponent in the middle. Meanwhile, the younger ram, aware that he would lose a painful full-scale struggle with either of his elders, is surreptitiously testing his budding strength and practicing his head-butting techniques. He delivers a sneaky but solid blow upon one of the occupied antagonists, then quickly retreats to avoid retaliation.

202

The midday sun is diffused by a tinkling haze of ice crystals as a band of bighorns feeds across a subalpine meadow. The group of two mature ewes, two yearling ewes, and three young rams left the main herd to escape the violent early December rutting activities. The temperature is −20 degrees, and the backs of the sheep are still matted with snow from the past night's storm.

Following pages: A sifting of hard snow crystals salts the face of a 300-pound bighorn ram during a subzero snowstorm. The chips in his huge horns are the scars of head-butting battles, and the horn tips are splintered from violent clashes with other big rams. The deep creases were formed each autumn when the horns temporarily stopped growing, and the segments between the creases represent the twelve years of his life. The annual growth gets smaller as the ram grows older, and this monarch is nearing retirement, for few live beyond the age of fourteen. The hairs caught in the broomed tips were pulled from other bighorns, perhaps a young ram rubbing its face against the old one's horns in a nonverbal display of subordination.

206

On the North American continent, mountain sheep are naturally separated into the thinhorn sheep inhabiting the Yukon and Alaska, bighorn sheep inhabiting the Rocky Mountains from southern Canada to Colorado, and desert sheep inhabiting the deserts from Utah to northern Mexico.

The Rocky Mountain bighorn sheep had its center of evolution in the mountains of the western United States and was one of the most plentiful animals found by early explorers. Modern man, the most ruthless predator of all, brought an end to the bighorn's mountain supremacy. Fencing and overgrazing of their grassland habitat, diseases introduced by domestic livestock, and indiscriminate killing overpowered the bighorn's natural defenses. Within one hundred years after the arrival of the white man in numbers to the Rocky Mountains, bighorns were reduced to some 10 percent of their original numbers. The survivors, restricted to fragmented remnants of their original range and caught in the crunch of inept land management, have been unable to rally. The beauty and promise of millions of years have been pitched to the edge of oblivion in an evolutionary instant.

Portrait of a novice mountaineer: A three-month-old bighorn lamb pauses uncertainly as it follows its mother through a difficult pass. The youngster's weight is rocked back on traction pads that grip the smooth, slanted surface. The mother called several times before the lamb made a desperation leap to the level below. Faced with being left behind, lambs inevitably find a way to negotiate difficult climbs or descents, thus becoming adept in the art of the impossible.

Beyond the Circle

TEXT BY LES LINE

IT IS DIFFICULT, perhaps impossible, to adequately describe the scenic grandeur of Alaska. Seventy-five years ago the famous cartographer of the American West, Henry Gannett, tried. The best he could offer was this advice to anyone who plans to visit Alaska: "If you are old go by all means; but if you are young stay away until you grow older. The scenery of Alaska is so much grander than anything else in the world that, once beheld, all other scenery becomes flat and insipid. It is not well to dull one's capacity for such enjoyment by seeing the finest first."

Yet in all of Alaska there is, today, only one national park, Mount McKinley, and two national monuments of note, Glacier Bay and Katmai. Their fame is fully justified, for the landforms they embrace are truly spectacular. But some of the finest of the finest scenery lies far to the north, above the arbitrary circumpolar line known as the Arctic Circle, and unprotected by any mandate of Congress.

In the minds of most people, the Arctic Circle defines the southern limits of the frigid north, a boundary that appears as rigid as one between two nations. In fact, it is an illusory line that is meaningful mainly to astronomers. It is at this latitude, 66½ degrees north, that for one twenty-four-hour day each summer the sun never sets, and for one twenty-four-hour day each winter the sun never rises. Scientists of other specialties, however, define the Arctic differently. To the botanist, for instance, it is the land that lies beyond the northern limit of tree growth. To the climatologist, the Arctic stops where the temperature, during the warmest month of the year, averages 50 degrees. Neither line follows a ruled, static course.

But no matter what one's persuasion, the Arctic is a rugged, beautiful, unspoiled region, and one of Earth's greatest sanctuaries for wildlife. Now Congress has an opportunity, one it will never have again, to set aside some wonderful pieces of the Arctic as national preserves under

Continued on page 231

212

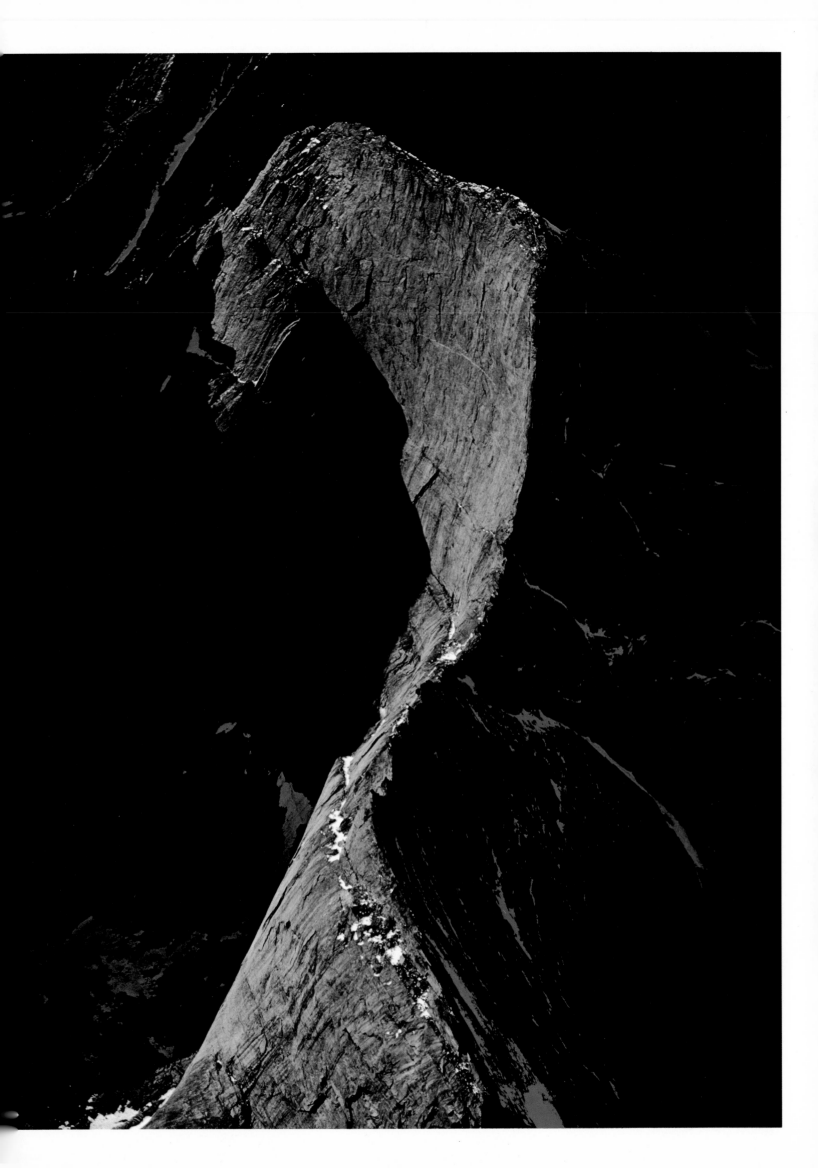

214

Arrigetch is an Eskimo word meaning "fingers extended," and Arrigetch Peaks is the very appropriate local name that Bob Marshall recorded in 1931 for this rugged group of mountains in the Brooks Range. A string of rock-basin lakes is nestled in the trough of a small glacier, and a stream tumbles down the path of the main valley glacier. Remnants of the glacial system survive at the foot of a granite ridge that has been plucked and polished into a series of monoliths —suggesting an Arctic Yosemite. There are thirty-six species of mammals in the Gates of the Arctic, including the majestic Dall sheep—as represented by this ram resting high on a bluff. (National Park Service photographs by Robert Belous)

Through the passes in the Gates of the Arctic pour thousands of caribou on their migrations each spring and fall. (Photograph by Steven C. Wilson)

The whiteness of the Arctic winter is reiterated in the plumage of the willow ptarmigan, searching for twigs . . .

. . . and in the coat of the snowshoe hare, feasting on needles of white spruce. (Photographs by Charles J. Ott)

Born of meltwater from glaciers high
on Mount Igikpak, loftiest peak in
the western Brooks Range, the great
Noatak River flows four hundred
and twenty-five miles to empty into
Kotzebue Sound. The river was first
explored in 1885 by an engineer from
the federal revenue cutter Corwin,
who described the "Grand Cañon
of the Noatak" where "perpendicular
walls rose hundreds of feet on either
side, seldom offering a foothold along
the bases." Here, in the Noatak's lower
canyon, white spruce grow in sheltered
areas beneath the bluffs. They are
at the northern limit of their range.
Indeed, a team of scientists studying
the Noatak ecosystem under the auspices
of the National Park System found it
to have one of the most varied floras
anywhere in the Arctic. The Noatak
usually runs clear in summer, but it
is clouded by sediment from heavy rains.
The land of the Noatak is the largest
river basin in America that is still
in its untouched state. It is the
major migration route for Alaska's
largest caribou herd, and its wealth
of wildlife includes grizzly bears,
Dall sheep, moose, wolves, and birds
of prey. (National Park Service
photograph by M. Woodbridge Williams)

the terms of the Alaska Native Claims Settlement Act. These photographs reveal the grandeur and the life of several of these areas—a Gates of the Arctic National Park, a Noatak National Arctic Range, an Arctic National Wildlife Range, a Selawik National Wildlife Refuge.

Henry Gannett suggested that one should see Alaska last. A modern-day government explorer, National Park Service planner John Kauffman, disagrees. "It is a stern country, of storm and cold and flood, but with summer warmth," he wrote. "You take this wilderness on its terms, not yours. It is no country for the casual. But it is readily enjoyable for men and women of all ages who are willing to accept its challenges, who cherish solitude and adventure."

A skilled hunter of the tundra, a red fox returns to its den with lunch —a ground squirrel. (National Park Service photograph by Robert Belous)

The Winter Marsh

PHOTOGRAPHS BY BILL RATCLIFFE

TEXT BY FRANKLIN RUSSELL

DURING THE night, the marsh froze hard. For days, the sun had fought against the freeze. It lit the cattail water gold and spun off six-sided diamonds of light from stalks of frost. Before the freeze, the marsh was a desperately tired place; summer wreckage everywhere, blackbird feathers forlorn in gray water, the bulrushes panzer-driven. The ice changed all that. It crept in, transparent and insecure at first, and muskrats smashed through it like tiny icebreakers. The ice persisted and the marsh was torn by a tumble of ducks reaching for black-water landings in thick early snow. The ice became blue and powerful; mergansers and golden-eyes lingered in narrowing pools until, one final frozen morning, the marsh was abandoned completely. Or so it seemed. Ice set the marsh in colored stone, held it still for the wild-watching eye, the silent artist, the motionless owl.

At first, the marsh's refrigerated death, its transfixion of all action, suggested a perfect wasteland. Nothing moved. But because there was time enough now to watch, and ample memory to stimulate, this was transparently a loose dictatorship. The ice had caught nothing unprepared. At night a rattle of dry stalks sounded, empty scabbards with the summer drawn out of them. In day, chickadees slid down the stalks and their feet encircled sleeping bees packed in vegetable corridors.

A sick duck occupied one last circle of black water for a day and was frozen in at the morning. Crayfish coiled in mud burrows, well-plugged, just below the line of any frost, and the frogs crouched deep in mud. A silent army of eggs lay stuck to stalks, and dropped in mud, and driven under stones, and drilled into stalks and wood and ancient leaves. The ice, by such prudent measurement, was more protector than jailer.

At the edge, and side, and bottom of the ice, a sleep of willing prisoners, of water bugs and beetles, of the nymphs of dragonfly and damselfly, of leeches gorged on autumn blood, of shellfish dormant, all growth stopped, at the threshold of immortal sleep.

The marsh was made, it seemed, of pitiless colored stone, but this was a device to trick the transitory watcher. It moved, violent, strong, and willful. It was pale today, full of gas bubbles,

"The marsh blended its shores—coasts of ice decorated with flowers of frost—into snow-drifted land. It moved, grew dense in silken mists, its outlines blurred and pale. The winter became an examination of how well the summer had provisioned the marsh."

234

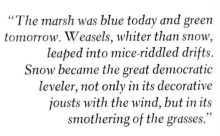

"The marsh was blue today and green tomorrow. Weasels, whiter than snow, leaped into mice-riddled drifts. Snow became the great democratic leveler, not only in its decorative jousts with the wind, but in its smothering of the grasses."

but purplish-black tomorrow with the bubbles driven out. It cracked and let up yellow fonts which froze into ice plants, leaves sprawling. It evaporated silently and drew up the stems of submerged plants and dead fish. It heaved, and up came unlucky frogs, not buried deep enough, and soon glassy-eyed under a weasel's foot. It moved and cattail jungles, still erect but shorn from their roots, moved obediently with it. A mountain wind spent days trying out different forms of snowdrifts, but gave up when its designs became too complex.

The marsh blended its shores—coasts of ice decorated with flowers of frost—into snowdrifted land. An ice window showed a distorted, familiar face, and a mink flickered away underwater. Shrews drilled ice tunnels, and moles, still digging with summer energy, dropped into the water and swam off. The silent blue ice revealed nothing of the torpedoing mammals and rudely awakened fish, the drowsy, jolted insects, the hunt continuing without rebate.

The marsh was blue today and green tomorrow, and saffron when the sun was kind. Precise crow footprints walked to a chiseled hole in the ice where an old dog muskrat, deep-frozen, was revisited daily. Weasels, whiter than snow, leaped into mice-riddled drifts; desperate tenants screamed injustice. Horned owls, arrogant in the temporary absence of crows, hunted in daylight. The rabbits were hidden and the foxes were visible, the chickadees here, the kingfishers gone. The ducks must be gone or dead, but two coots defied reason and clumsy-footed across a blue arena in the dead cattail crowd. Newts, sluggish among under-ice debris, ate comatose insects and kept growing. On the coldest day of the year, they shed their skins as though midsummer reigned.

Snow became the great democratic leveler, not only in its decorative jousts with the wind, but in its smothering of the grasses, its clothing of the marsh in wool. It fell against the sun which glittered among its crystals and petrified its purpose, revealed sexagonal bodies, and countless different flake forms radiating from heartlike cores. Meanwhile, a steady growth of frost flowers decorated every edge of the marsh.

The crystals settled and trapped the warmth of both atmosphere and earth. The snow might

236

thaw, then freeze into a crust, but the warm air remained well trapped. Inside the snow, it was possible to laugh at the winter.

The marsh moved, grew tense in silken mists, its outlines blurred and pale, and a barred owl's vicious scream bit into partridge ears. Quail felt the same tension, stirred uneasily in their snowdrift, and at dawn, one stayed behind to die. The cold toughened. Snow tracks converged, mink footprints galloped between teardrops of blood. A rabbit spoor ended in one disheveled question mark. A muskrat's tail was welded against a cattail stalk, but its body was gone; a drag mark moved into reed forests.

The winter became an examination of how well the summer had provisioned the marsh. Muskrats chewed the underparts of the cattails and bulrushes, and fish breathed the last of the oxygen collected in bottom hollows. Foxes and mink cut down the dazed victims of the deepest freezes. If there were no thaw, or too much snow, or not enough, a plague of death must strike the marsh.

The marsh breathed deeply. It had manufactured oxygen in that grand corrupt time of high summer when, richly decomposing, its plants had liberated it everywhere. Air-laden streams and runoffs brought the precious stuff in from all sides. The V-shaped swim tracks of muskrats cut through water alive with oxygen; ducks and bitterns and ovenbirds and redwings shouted uproarious paeans to it. The snapping turtles and minnows and others thrived in its generous presence when it was inconceivable that the marsh could ever be frozen into lunar suffocation.

The ice grew through the coldest month. Only dim light reached photosynthetic factories and production ebbed low. Ice thickened in upland streams. A trickle of water slid into the marsh. In resealed caverns, desperate fish gulped and choked and suffocated and wise crows waited for the bodies to be sent through the ice to them.

The crisis passed; the marsh thawed. In a long hiss of expiring flame, the sun dropped from the day's work of melting the ice. It colored the marsh pink, dilute blood, and that was apt. The spring flowed secretively in submarine arteries as a blizzard smothered the work of the sun. It flowed in the bodies of young salamanders growing now as the ice thickened again. It flowed and fattened young tadpoles. Aquatic insects kicked themselves forward to the tunes of a bitter cold gale, and grew with tiny mites and crustaceans. The marsh was stealing time from the season; every muscle in its still-frozen body toned for a rush into spring. The crayfish were paired in the gloom, the fairy shrimps and water fleas packed brood pouches with embryos. Another month of ice could not halt or slow the spring of the frozen marsh.

*"Early in the morning we found meltwater
from the previous day frozen over
mats of blueberry and pine needles."*

Ice Shall Reign

PHOTOGRAPHS AND TEXT BY DAVID CAVAGNARO

THERE WAS a time, just a moment ago on the clockface of the universe, when the rock of the Earth was fluid. There will also be a time, in the dying days of our sun, when the last internal boiling of Earth rock will be stilled, and the frozen planet will spin on through the void of space until the fires of another star embrace it.

The universe has a sense of time far greater than our own. Heat and cold have their particular times and places, their own enormous rhythms. The Earth lies somewhere in transition within this larger cycle; solid, liquid, and gas coexist. This is a brief and beautiful instant, and we are lucky enough to be alive to witness it.

No compound on Earth illustrates this passing fragment of planetary history more vividly than water. Alone among all the components of our planet, water exists in all three states within prevailing temperatures. It fills the oceans and rivers, covers mountain peaks and the poles, permeates the atmosphere, and is locked within the crust itself. Water controls the weather and shapes the land. Most amazing of all, water is the principal ingredient of life, that peculiar borderline creation, fleeting magic of the transition between fire and ice. Life was born of water in its liquid form, and life will probably die by water when the Earth freezes over.

Universe time lies quite beyond our grasp. We, as the predominant and dominating form of life on Earth, are correct in worrying more about the current health of our planet than about its eventual ice-bound death a few billion years from now, lest we bring about the end of life ourselves much sooner. But we are fortunate indeed to have within our own experience a much smaller sample of the fire and ice cycle of the universe. With each revolution of the Earth around the sun we witness the rhythms of the seasons, and with each rotation we know the warmth of day and the coolness of night.

During the middle of October my wife, Maggie, and I accompanied photographer Ernest Braun and his son, Jonathan, on a backpack trip to Iceberg Lake in the Minarets Wilderness of the Sierra Nevada. Winter was pending. As we approached the lake, clouds hung low among the

Continued on page 245

240

"The ice patterns
on rivulet and puddle
were believable only
because they existed.
When the sun appeared
briefly through clouds,
the ice melted a little
around the edges, but
the ice always gained.
Beneath a small fall,
clumps of rushes with
flowers still in bud
were encased in sheaths
of frozen spray."

Minarets, ensuring the survival of the last icebergs that had escaped, at least in part, the work of the summer sun in the high country.

An icy wind blew the first light snow flurry out of the north. For two days the temperature lingered at twelve degrees and below. Iceberg Lake began to freeze over. Now and then a new berg would fall from the glacier at the head of the lake, and all the surface ice would crack and tinkle like a million windows breaking. We were on the fringe of winter, drawn by the thrill of impending change and the exhilaration of the unexpected. A blizzard was due; we might as well have been reaching far over the edge of a great precipice for a better view of a strange element, the air itself, for which birds, not man, are adapted. Not another soul roamed the area except for a very brave (or crazy) artist packing a five-foot easel.

It had been a short and tough summer for the alpine plants. Many flowers still struggling to bloom were bruised and limp from the cold. Early in the morning we found meltwater from the previous day frozen over mats of blueberry and pine needles, and beneath a small waterfall that defied the approaching grip of winter, clumps of rushes with their flowers still in bud were encased in sheaths of frozen spray.

The ice patterns on rivulet and meadow puddle were believable only because they existed. Like growing tissue covering a wound, ice crystals closed in upon the last bits of runoff, winter's attempt to heal the destruction summer had wrought among its splendid ice fabric.

When the sun appeared briefly through the clouds, the ice melted a little around the edges, the last insect larvae of the season came sluggishly to life in the icy pools, but the ice always gained because summer was over.

The Earth knows many rhythms, among them night and day, summer and winter, glacial and interglacial. Before the uplift of the Sierra, this place was another range of mountains, and before that a shallow sea. Throughout all these changes, life has adjusted. The flowers, some wilted as they were by frost, would survive; ground squirrels were preparing for hibernation; summer birds were on their way to warmer places, and the timberline pines gripped the granite and faced squarely into the teeth of winter.

For our own part, we knew that we were visitors here. Our roots were in other, milder places. The Minarets' freeze was only a moment in our lives, yet on another scale man as a species will occupy only a moment in the life of the Earth. We are visitors on the planet as well. The larger balance of fire and ice transcends us. We had been given a tiny glimpse of the giant physical powers that prevail in the universe, and we felt relieved by the reminder that man, for all his own striving after power, will not be able to control all of them.

For 54 years, AUDUBON and its predecessor, BIRD-LORE, were digest-sized. The green-tinted, Gothic-lettered cover— unchanged since 1899—was finally transformed by young Roger Tory Peterson in 1935. The name was changed in 1941. (Photographs by Alan Fontaine)

Postscript:
The Evolution of a Magazine

BY ROGER TORY PETERSON

IT IS not generally known that way back in 1887 and 1888 there was a monthly publication named *The Audubon Magazine*. Devoted to songbird protection, it was "published in the interests of the [Audubon] Society" by the Forest and Stream Publishing Company of New York. The annual subscription was fifty cents, and a single copy cost only six cents.

Dressed in a gray cover with a steel engraving of John James Audubon, it ran such articles as "How I Learned to Love and Not to Kill," "A Plea for Our Birds," "Bluebird Dick," "Wrens in a Coffee Pot," and "Reintroduction of Feather Millinery." And a fantasy series for children entitled "Charley's Wonderful Journeys." The passenger pigeon and the Carolina parakeet were not yet extinct—not by some years. But the magazine and the society to whom it was dedicated soon would be; America, unaware that it was rapidly losing its wild heritage, was not yet ready.

With the urgent revival of the Audubon movement under a different aegis eight years later, it became evident that an official organ was needed if the fledgling conglomerate of local societies was to survive. The magazine that filled the niche was *Bird-Lore*, financed and launched in February 1899 by Frank M. Chap-

man of the American Museum of Natural History, who became the dean of American ornithologists.

"A Bird in the Bush Is Worth Two in the Hand" was the motto that appeared on the *Bird-Lore* masthead. The first nine bimonthly issues were, with Mrs. Chapman's assistance, sent out from their home at Englewood, New Jersey; but the nights were scarcely long enough for the labor involved. In August 1900 the operation was moved to Harrisburg, Pennsylvania, where the Macmillan Company, the publisher, and later J. Horace McFarland Company, the printer, took on the task of mailing and distribution. Price in those days? One dollar a year, twenty cents per copy.

In his "prospectus" setting forth the aims of the magazine, Chapman had stated that every prominent writer on birds promised to contribute to *Bird-Lore*. Before the second year was up he made good the promises with articles by William Dutcher, A. K. Fisher, Ernest Thompson Seton, William Beebe, Robert Ridgway, Mabel Osgood Wright, Florence A. Merriam, J. A. Allen, William Brewster, Witmer Stone, Bradford Torrey, Otto Widmann, and many others. The very first article was "In Warbler Time," by the "sage of Slab Sides," John Burroughs. Three years later, Ernest Thompson Seton, writing on the

recognition marks of birds, presented the first overhead flight patterns of the birds of prey.

The annual Christmas bird count was born the second year. In *Bird-Lore* for December 1900, Frank Chapman wrote: "It is not many years ago that sportsmen were accustomed to meet on Christmas Day, 'choose sides,' and then . . . hie themselves to the fields and woods on the cheerful mission of killing practically everything in fur or feathers that crossed their path. . . . Now *Bird-Lore* proposes a new kind of Christmas side hunt in the form of a Christmas bird census, and we hope that all our readers who have the opportunity will aid in making it a success by spending a portion of Christmas Day with the birds and sending a report of their 'hunt' to *Bird-Lore* before they retire that night."

A total of twenty-seven people from twenty-five localities took part in the first Christmas count in 1900. In succeeding years the count became so popular that by 1971 the number of participants grew from that small nucleus to 18,798, representing 963 areas. That report (like the preceding sixteen reports, it was edited by Allan Cruickshank) is exactly one inch thick. Analyzed intelligently, these counts, as well as the seasonal reports of field observations in various parts of the country that have been published since 1918 under the title of "The Season," tell us much about the population trends of birds. From the conservationist's point of view, shifts in bird populations are excellent indicators of the environment—a sort of "ecological litmus paper."

"The Season," the "Breeding-Bird Census" (started in 1937), and the Christmas count, all of which were originally included among the pages of the magazine, were separated in 1939 to be published independently as *Audubon Field Notes* (now *American Birds* under the editorship of Robert Arbib).

Perhaps the most extraordinary single feature of the magazine during the long Chapman era, a contribution of lasting importance, was the series of bird portraits by Louis Agassiz Fuertes, George Miksch Sutton, Allan Brooks, R. Bruce Horsfall, and others. For many years, a frontispiece in full color graced each issue. These portraits were accompanied by migration data organized by Frederick Lincoln of the U.S. Biological Survey, and notes on the plumages of North American birds by Chapman himself. The series of

hummingbird plates by the great Fuertes is especially vivid in my memory; never have these gemlike birds been painted more brilliantly.

It was quite true that some of the material in earlier issues verged on "corn," especially the verse. The funniest, undoubtedly, were several stanzas by T. Gilbert Pearson that began: "Which would you choose for life's short whirl, the girl with the gun, or the camera girl?" The poem was illustrated by a wash drawing of a young lady with a box camera and another with a shotgun, shooting quail.

When John Baker took the helm of the National Association of Audubon Societies on November 1, 1934 (the same day that I joined the staff), he had many plans for revitalizing a much-weakened organization. Membership had declined drastically. This was partly due to the Depression, but also to the critical attacks on the integrity of the Audubon leadership by Mrs. Rosalie Edge.

At that time Frank Chapman was finding his magazine more and more of a burden, and rumor had it that when he was away on expeditions he turned over the entire editorial responsibility to his secretary. Some years earlier T. Gilbert Pearson, then president of the National Association of Audubon Societies, complaining that conservation items and news of affiliated clubs were relegated to small print toward the rear of the magazine, had tried to buy *Bird-Lore* from Dr. Chapman but had been turned down. Now the time was more propitious. With his usual directness, John Baker suggested to Dr. Chapman that the association should acquire the magazine and assume full editorial and publishing responsibility. Dr. Chapman, who had edited his journal for thirty-six years while writing fifteen books and more than two hundred technical papers—and also while building up the most extensive ornithological collection to be found in any museum in the world—was now only too glad to be relieved of a responsibility that had become an unrelenting chore. In the January–February 1935 issue of *Bird-Lore* it was announced that "Dr. Chapman has sold to the association all right, title, and interest to and in the assets and income of *Bird-Lore.*"

At the time of the transferal Dr. Chapman wrote: "Who can estimate the value to birds, bird-lovers, and bird-lore of *Bird-Lore*'s 14,000 or more pages and nearly 5,000 illustrations?" Never in the history of popular ornithology had there been anything like it. It was a voice that had helped save the egrets and other plume birds, a voice that was the first to speak out in many a conservation crisis.

The number of subscribers to *Bird-Lore* at the time of the takeover was in the neighborhood of five thousand. Inflation had not affected the subscription rate much. It had been held to $1.50 a year, less than the price of a single copy today. The cover of the magazine with its old-fashioned design and Gothic lettering had the look of a religious journal. The new editor, William Vogt, soon changed that. It fell to me to design the new cover. Although I had once attended lectures by Frederic W. Goudy at the Art Students League, I was not a skilled calligrapher; nevertheless, I came up with some new lettering. This face-lifting incorporated a painting by myself of a rough-legged hawk on the wing over a wintry landscape as viewed from above by another hawk. It came off fairly well, and I was wowed when I received a letter from the legendary Arthur Cleveland Bent of *Life Histories* fame, stating that it was one of the best bird paintings he had ever seen. That may have been an overstatement. In contrast, the next cover, five shovelers springing from the reeds, was a disaster. In planning covers we were limited to two ink impressions, black and one color. The color could either be used in the title, on the bird (as the red on a rose-breasted grosbeak), or as a background tone (green water, blue sky, etc.). The limitations we worked under in those days!

When William Vogt was "discovered" by John Baker he was acting as curator of the Jones Beach State Bird Sanctuary on Long Island. Before that he had been a drama critic, an editor for the New York Academy of Sciences, and a writer of a nature column that appeared in five Westchester County newspapers. I was the one who had introduced him to birds. After a number of field trips he suggested that I put my system of field identification into book form. I did, dedicating the compact little volume to him. Thus was my first field guide born.

Bill, badly crippled by polio suffered at a summer camp when he was a teen-ager, was a born crusader and a very articulate critic. For each issue of the magazine he wrote a hard-hitting editorial on some conservation crisis, a practice that was discontinued when he left the society. Of course, not all of the magazine's succeeding editors have had Bill's insights.

During Vogt's editorship the society gained stature; no longer was it identified mainly with the "white-breasted nuthatch type of birdwatcher." Under the guidance of John Baker as president and Vogt as editor, the Audubon movement shifted gears, and led the

way to an environmental awareness that has finally reached the grassroots throughout the country. Those early articles, by such prophets as Paul Errington, Aldo Leopold, Robert Cushman Murphy, and Paul Sears, have since been elaborated on by a thousand other writers, but some of the pioneer essays, written three decades ago, remain classics.

In a conservation magazine there is always danger of overpessimism, and some subscribers during Vogt's tenure complained of this, as indeed subscribers do today. They wanted to be diverted, instructed, and entertained, not preached to. One issue, I recall, was undeniably depressing. There was an essay by Errington about the suicidal tendencies of muskrats under population pressures; another described the difficulties of winter survival of quail; a third dealt with the ecological effects of poisons—the entire issue reeked of death and destruction.

Most articles in those days were solicited by the editor (as they still are today), but the authors were not paid. There was no money in the budget for this purpose, and therefore professional writers were ruled out. Most academic men cannot write a popular article without resorting to the technical jargon and clichés of their discipline. However, a few such as Paul Errington, Aldo Leopold, Robert Cushman Murphy, and George Miksch Sutton were able to express themselves in lucid, often beautiful prose. Outstanding also were the book reviews that under later editors slipped badly or were abandoned entirely.

It was inevitable that Bill Vogt, stormy petrel that he was, should have differences with such a dominant personality as John Baker. They did not always see eye to eye, and it seemed to rebel Vogt that Baker should be deposed. Fomenting an office revolt in 1938, he pulled most of the other members of the staff into his camp. Although Bill scored some valid points, the board of directors decided to stand behind John Baker and that instead, Bill Vogt must go.

Robert Cushman Murphy, much impressed by Bill's scholarship, recommended him to the Peruvian Guano Administration to undertake a study of the population dynamics of the guanay cormorant and other guano-producing birds of the Chinchas, those fabulous bird islands in the Humboldt Current off the coast of Peru. So Bill, a few weeks after he relinquished the editor's chair, sailed for Peru, "to help increase the increment of the excrement," as he put it.

It was during these three years, while observing the guanays and piqueros, which are subject to extraordinary population fluctuations, that Bill began to think about human overpopulation problems, especially in South America. Fluent in Spanish, he involved himself in Latin American conservation, and at the end of his Peruvian stint was employed by the Pan American Union as chief of its conservation department. While in Washington he wrote his great classic, *Road to Survival*, a neo-Malthusian epic that became a national best seller, highly praised and roundly damned. *Time* gave four pages to a critical review, labeling it alarmist. Today *Time* itself, aware of the population crisis, espouses the very ideas it once condemned.

The recent breakthrough in environmental awareness on the part of the general public is often credited to Rachel Carson's *Silent Spring* but it can also be traced back to Vogt's *Road to Survival*, and even further to some of the articles and editorials under Vogt's editorship in *Bird-Lore*.

At Vogt's departure for Lima, Peru, in January 1939, the society was faced with the problem of finding a new editor. George Miksch Sutton was considered and was called in to confer with John Baker, Guy Emerson, and Robert Cushman Murphy. He was much tempted by the opportunity offered him but too deeply involved in his life as curator of birds at Cornell University to make the switch. Instead, he agreed to become a contributing editor.

Meanwhile Margaret (Peggy) Brooks, the society's librarian, kept the publication going on the strength of her earlier newspaper experience and her apprenticeship to Vogt while he was editor. For three more years she maintained the critical standards set by Vogt, all the while presiding over the library, a work load that seems almost inconceivable today.

It was during her editorship that the magazine underwent another bit of major facial surgery because the title, *Bird-Lore*, had ceased to symbolize adequately the society's work. With the January 1941 issue it appeared as *Audubon Magazine*. Twenty years later, in 1961, the word *Magazine* was dropped; it is now simply *Audubon*.

During the Vogt-Brooks years, a young man, Joseph J. Hickey, a troubleshooter for Consolidated Edison Company, often took two-hour lunch breaks in the Audubon library to pursue his real passion, ornithological research—or was it Peggy Brooks? When he decided to leave New York to seek a master's degree at the University of Wisconsin, Peggy went with him as Mrs. Hickey. Joe, who initiated the "Breeding-Bird Census" in *Bird-Lore* in 1937, is now professor of wildlife ecology at Madison and is currently president of the American Ornithologists' Union. Peggy, in her spare time, still edits technical publications on ornithology.

In midsummer of 1943, it was announced that Eleanor Anthony King, a journalist and writer of children's nature books, had been appointed as the new editor. A pleasant, friendly woman in her early forties, Eleanor had been quietly editing the magazine for a year before she allowed her name to be put on the masthead. To use the words of one of her favorite contributors, Alan Devoe, "She was a person of immense life-zest and high spirits, a vitality constantly spilling over into laughter, antical jokes, and an uproarious contempt for the narrowly conventional, the pretentious, the stuffy."

She once said that she would like the magazine to have "the same general appeal as *Collier's*." It became a point of debate around the office as to which was more effective in meeting the conservation challenge—a superficial mass audience, or a smaller, more sophisticated readership. As if in answer to this, *Collier's* died of strangulation less than a decade later. Nevertheless, though she lacked the biological insights of William Vogt, Eleanor King broadened the scope of the magazine, transforming it into a more vividly illustrated and more lively periodical. Stimulating, browbeating, and coaxing some of the best naturalists in the country—luminaries such as Louis J. Halle, Richard H. Pough, Ludlow Griscom, Sally Carrighar, Edwin Way Teale—to write articles, Eleanor King edited the magazine for seven years, her energy and devotion seeing it through the difficult war period. At age forty-seven she succumbed to cancer.

At this point, John K. Terres, who had assisted Miss King with the magazine during her illness, took over as editor under Kenneth D. Morrison, the society's public affairs director. In 1956 Terres became editor in title as well as in fact. For eleven years he ran the magazine, longer than any editor since Chapman. He was also one of the magazine's most eminently qualified editors, with prior experience as a biologist in the employ of the Soil Conservation Service and with a distinguished record of publication, both popular and technical.

The magazine continued to evolve and to improve year by year under the skilled hand of John Terres, who was far more than a routine editor. He was an observant naturalist in the true sense, and his enthusiasm for his subject spilled over in numerous footnotes. Footnotes are a device more often used by British nature writers than by Americans, and John Baker may have thought that Terres resorted to them too often. However, I liked them because I felt better informed about details that were not developed by the authors.

John, who wished to resume his own creative writing, found it difficult to do so while tailoring other people's pieces. In 1961 he relinquished the editor's chair to return to the woods and fields. He was one of the best editors the magazine has ever had, and it was no surprise when he received the prestigious John Burroughs medal in 1971 for his book *From Laurel Hill to Siler's Bog: The Walking Adventures of a Naturalist.*

John Vosburgh, whose brother, Frederick Vosburgh, was editor of *National Geographic,* succeeded John Terres, bringing to the magazine the same dedication to wildlife conservation as his predecessor. Just turning fifty, he was a journalist with a background as a sportswriter for the *Washington Post* and assistant Sunday editor for the *Miami Herald.* As a nature columnist, he strongly supported the establishment of a refuge for the Key deer, adoption of the final boundaries of Everglades National Park, protection of the Everglade kite, and other Audubon projects. A well-informed and active conservationist, Vosburgh was editor for five years and is now with the National Park Service.

Although there had been a single cover in full color back in 1952, a photograph of a whooping crane by Don Bleitz in the January–February issue, color covers did not become a standard feature of *Audubon* until the September–October 1961 issue, when Frederick Kent Truslow's Kodachrome of a pair of bald eagles was reproduced.

Our present editor, Les Line, was bold enough to venture into color extensively through the magazine, resulting in what is now not only the most beautiful natural history magazine in America, but in the world. Indeed, many people are now calling it the most beautiful magazine *of any sort* in the English language. I agree. This quality has had the inevitable effect of sending Audubon membership rocketing. Editorially, the magazine is as hard-hitting as it was during the days of Bill Vogt, but it is undoubtedly much more influential because of its vastly greater circulation, a figure expected to exceed 210,000 with this historic issue.

When I first met the young genius who now sits in the editor's chair, I had no idea that he would someday transform the magazine with which I had been so closely associated. It was a bitterly cold day on the marshes bordering Lake Huron. At that time he was employed by the *Midland* (Michigan) *Daily News,* and I recall that we talked mostly about cameras and about birds of prey.

Les has surrounded himself with a highly skilled staff, and under his guiding hand the magazine has been paying competitive fees for articles, attracting some of the best writers in the field. Although this emphasis on good writing is not lost on the readership, the visual metamorphosis has caused the greatest comment.

In his "prospectus" setting forth the aims of *Bird-Lore,* Chapman had promised that the magazine would print the best photographs of living birds ever published in this country. But photographs of wild birds were hard to come by. Edwin Way Teale, running through the issues for 1900–1901, counted eighty-eight photographs of stuffed birds to only seventy-six of living birds, and several of the latter were taken in zoos.

But if we look back to Chapman's day we realize that the magazine has spanned the entire history of bird photography. Frank Chapman himself is said to have taken the first photograph of a wild bird as early as 1888. In those days, of course, there were no such things as color film, strobe lights, exposure meters, miniature cameras, automatic lenses, or any of the other refinements that we now take for granted.

Way back in the first issue, Thomas S. Roberts, who was to become the author of the ponderous two-volume *Birds of Minnesota,* wrote a pioneer article on photography as a tool in bird study. During the early years of *Bird-Lore,* Chapman dominated the field of bird photography. Then came Arthur A. Allen of Cornell, who contributed a section for young readers. During the precolor period and for a decade after the society took over the magazine, the superb portraits by Allan Cruickshank literally swamped out the others, and we can rightfully say that he became the dean of black-and-white bird photographers in America—as Eric Hosking is in England.

Today, with the extensive use of color, *Audubon* leans more toward the extraordinarily detailed salon compositions of Eliot Porter and the mood-conveying, selective focus work of Pat Caulfield, David Cavagnaro, and others.

In the first issues of *Bird-Lore* the advertisements were solely for new bird books; later, cameras were offered ($5.00 to $10.00!). Today birdwatching is big business, and I need not go into all the products advertised. However, one development stands out— nature tourism. Guided tours, literally to the ends of the Earth, and often led by competent naturalists, are now offered. The amateur birdwatcher and naturalist roams the world, and *Audubon,* cognizant of this development, now looks beyond our borders to the other continents and to the seven seas of our small blue planet, whose future has become our concern.

Contributors

The Photographers

DENNIS BROKAW scours the North American continent's rugged western shoreline and its coastal forests for the beautiful and perceptive details that are his photographic hallmark. Many of these pictures appear in his book THE PACIFIC SHORE: MEETING PLACE OF MAN AND NATURE. He abandoned a career as a flight-test engineer for the life of a nature photographer, and his work is featured regularly in natural history and photographic journals, in books, and in exhibits.

PATRICIA CAULFIELD was executive editor of *Modern Photography* when, in 1967, she decided to devote all her time to nature photography. Much of that time, over the past seven years, has been spent documenting in her distinctive way the fast-vanishing wild places of Florida. She created the Sierra Club's large-format book on THE EVERGLADES; contributed the photographs for THE OKEFENOKEE SWAMP in the Time–Life book series on The American Wilderness; and photographed the beautiful Oklawaha River, being ruined by a barge canal across the state, for *Audubon*. And she has returned to her home state of Iowa many times to portray the last remnants of the original prairie.

DAVID CAVAGNARO is resident biologist at Audubon Canyon Ranch, a wildlife sanctuary and nature education center, located north of San Francisco, that includes a heronry in the treetops of a redwood grove. An entomologist, he has joined scientific expeditions to India, Southeast Asia, Australia, Central and South America, Mexico, the Galápagos Islands. His photographic trademark is the extreme closeup view of nature—of insects, spider webs, drops of dew, frost, wildflower seeds, abstract patterns. He is also a poetic writer who provided both the words and pictures for his book THIS LIVING EARTH.

JOHN DEEKS, as his contribution to this anthology suggests, is fascinated by weather—especially by storms. And there is no better place to observe these phenomena than atop a mountain, from the vantage point of a fire tower. Thus he chooses to spend his summers as a lookout for the U.S. Forest Service, guarding the trees from conflagration—and focusing his camera on passing clouds, on the fireworks of thunderstorms, on sunsets and moonrises. During the off-season, he experiments with making high-quality black-and-white photographic prints using techniques long ago abandoned as too time-consuming.